AGRI-FOOD COMMODITY CHAINS
AND GLOBALISING NETWORKS

Agri-Food Commodity Chains and Globalising Networks

Edited by
CHRISTINA STRINGER
and
RICHARD LE HERON
The University of Auckland, New Zealand

Routledge
Taylor & Francis Group

LONDON AND NEW YORK

First published 2008 by Ashgate Publishing

2 Park Square, Milton Park, Abingdon, Oxon OX14 4RN
711 Third Avenue, New York, NY 10017, USA

Routledge is an imprint of the Taylor & Francis Group, an informa business

First issued in paperback 2016

British Library Cataloguing in Publication Data
Agri-food commodity chains and globalising networks. - (The
    dynamics of economic space)
    1. Produce trade 2. Globalization - Economic aspects
    3. Produce trade - New Zealand 4. Globalization - Economic
    aspects - New Zealand
    I. Stringer, Christina II. Le Heron, Richard B.
    338.1'9

Library of Congress Cataloging-in-Publication Data
Stringer, Christina.
Agri-food commodity chains and globalising networks / by Christina Stringer and Richard Le Heron.
        p. cm. -- (Dynamics of economic space)
    Includes index.
    ISBN 978-0-7546-7336-1
    1. Agricultural industries. 2. Commodity exchanges. 3. Globalization. I. Le Heron, Richard B. II. Title. III. Series.

    HD9000.4.S77 2008
    338.1--dc22

                                                            2008002423

ISBN 13: 978-0-7546-7336-1 (hbk)
ISBN 13: 978-1-138-25215-8 (pbk)

Contents

List of Figures

List of Tables

List of Contributors

Paula Blackett is a social scientist at AgResearch, Ruakura Research Campus, Hamilton, New Zealand. Her area of research can be broadly defined as around human dimensions of environmental issues. Email: Paula.Blackett@agresearch.co.nz

Hugh Campbell is Associate Professor of Social Anthropology at the University of Otago with a research interest in agriculture and sustainability. He is the Foundation Director of the Centre for the Study of Agriculture, Food and Environment, a Centre which specialises in research into agriculture working at the interface between the social and environmental sciences (see www.csafe.org.nz). Email: hugh.campbell@otago.ac.nz

Michael C. Carroll is the Director of BGSU's Center for Regional Development and Associate Professor of Economics. Dr Carroll's business and economics career dates from 1982 and includes service as an assistant professor of economics, corporate controller, operations manager, and a corporate presidency. His research interests focus on regional economic development strategies and social economics. Email: mcarrol@bgnet.bgsu.edu

Stuart Gray is currently General Manager of Global Trade NZ at Fonterra Co-operative Ltd and has worked in the dairy industry for more than 20 years, based in New Zealand, Australia and the Middle East. He holds an M.A. in Geography from The University of Auckland. Email: stuart.gray@fonterra.com

Lesley Hunt is a research fellow at the Agricultural Economics Research Unit at Lincoln University (New Zealand) with interests in the identities and symbolic meanings employed by farm households. Email: Lesley.hunt@lincoln.ac.nz

Brian Ilbery is Professor of Rural Studies at the Countryside and Community Research Institute, University of Gloucestershire. Brian has research interests in agricultural change and policy, rural development, local and regional food supply systems and alternative farm enterprises. He is currently working on major research projects on risk and plant disease management and local/national organic food markets. Email: bilbery@glos.ac.uk

Mairi Jay is a Senior Lecturer at the University of Waikato, New Zealand. She is Acting Director of the Environmental Planning programme within the Department of Geography, Tourism and Environmental Planning. Her research interests include sustainable land management and conservation of biodiversity in production landscapes. Email: mairij@waikato.ac.nz

Elmar Kulke is Professor of Economic Geography at Humboldt-University, Berlin. He is also President of the German Society of Geographers; his fields of research include commercial geography, locational systems, Central Europe, Southeast Asia. Email: elmar.kulke@geo.hu-berlin.de

Richard Le Heron is Professor of Geography at The University of Auckland and has longstanding research interests focusing on the transformations in economic activities under neo-liberalising conditions. Over the past decade he has examined issues of regulation, governance and governmentality with reference to supply chain re-alignment in the agri-food sector and fisheries management and the tensions between growth and innovation policies and policies relating to sustainability in urban and rural contexts. Email: r.leheron@auckland.ac.nz

Nicolas Lewis is a Senior Lecturer in Geography at The University of Auckland and works with post-structural political economy approaches. He has research interests in industry institutions, neoliberal political projects, and geographies of education. Nick has followed closely the development of the New Zealand and global wine industries, and is currently researching the proliferation of nation branding. Email: n.lewis@auckland.ac.nz

Charles Mather works in the School of Geography, Archaeology and Environmental Studies, University of the Witwatersrand, Johannesburg. His research focuses on the transformation of Southern Africa's agri-food system. Email: charles.mather@wits.ac.za

Damian Maye is a senior research fellow at the Countryside and Community Research Institute, Cheltenham. He is a rural geographer whose research interests lie in agricultural restructuring, food supply chains and rural development. Damian is particularly interested in the growth of local/alternative food networks, the geography of micro-brewing, CAP reform and farm tenancy and, more recently, issues associated with biosecurity and risk management in the countryside. Email: dmaye@glos.ac.uk

Jeffrey Neilson is a research fellow in Geography at the University of Sydney. His research interests include agricultural change, rural development and commodity trade across South and Southeast Asia. Email: jneilson@geosci.usyd.edu.au

Jerry Patchell is Associate Professor, Social Science, Hong Kong University of Science and Technology, Hong Kong. His research interests include interfirm and institutional cooperation in various spatial, industrial and environmental contexts. Email: sopatch@ust.hk

Eric Pawson is Professor of Geography, University of Canterbury. He has undertaken a range of work in recent years on New Zealand's environmental histories with Professor Tom Brooking of the University of Otago. They have been co-leaders of the funded 'Empires of Grass' project from which both his chapter, and that by Dr Vaughan Wood, in this volume are drawn. Email: eric.pawson@canterbury.ac.nz

Bill Pritchard is an Associate Professor in Economic Geography at the University of Sydney with interests in the geographies of global change in agriculture, food and rural places. Email: b.pritchard@geosci.usyd.edu.au

Eugene Rees completed his PhD, on which his chapter is based, at The University of Auckland in 2005. He is now a socio-economic analyst at the New Zealand Ministry of Fisheries. His research interests include fisheries supply chain alignment and individual firm behaviour. The views expressed in the chapter are those of the author and do not necessarily represent those of the Ministry of Fisheries. Email: Eugene.Rees@fish.govt.nz

Neil Reid is Associate Professor of Geography and Planning and Director of the Urban Affairs Center at the University of Toledo. His current research interests are in the theory and practice of industrial clusters. Email: neil.reid@utoledo.edu

Michael Roche is Professor of Geography in the School of People Environment and Planning at Massey University, Palmerston North, New Zealand. His research interests include the historical geography and contemporary geography of agriculture and forestry systems in New Zealand. Email: M.M.Roche@massey.ac.nz

Christopher Rosin is a research fellow examining issues of agricultural sustainability at the Centre for the Study of Agriculture, Food and Environment (CSAFE) at the University of Otago in Dunedin, New Zealand. Email: chris.rosin@stonebow.otago.ac.nz

Andreas Stamm is working at the German Development Institute (DIE), a think tank for Germany's development policy. His research interests are related to private sector development promotion and the possibilities of developing countries to adapt to the conditions of a globalising knowledge society. Email: andreas.stamm@die-gdi.de

Christina Stringer is a Senior Lecturer in the Department of International Business, at The University of Auckland. Her research interests include agri-food and forestry industries and in particular she is interested in governance relationships within the global value chain in the form of benchmarking and certification standards. Email: c.stringer@auckland.ac.nz

Christine Tamásy is Heisenberg Research Fellow at the School of Geography, Geology and Environmental Science, The University of Auckland, New Zealand. At present she is involved in a number of projects concerning globalisation, migration and entrepreneurship in the Australasian context. Email: c.tamasy@auckland.ac.nz

Vaughan Wood is a historical geographer, specialising in agricultural history, based in the Geography Department, University of Canterbury, New Zealand. His chapter is based on postdoctoral research undertaken for the 'Empires of Grass' project, which is co-led by Professor Eric Pawson of the University of Canterbury (who also has a chapter in this volume) and Professor Tom Brooking of the University of Otago. Email: gareth.wood@canterbury.ac.nz

Referees

The editors would like to thank all the reviewers who reviewed individual chapters. The refereeing process consisted of two parts: an assessment by a referee selected from conference participants and an international referee in the agri-food field. The reviewing process was anonymous. Special thanks, in particular, goes to the external referees who generously gave of their time: Montserrat Palares-Barbera (Universitat Autònoma de Barcelona); David Barling (City University, London); Arthur Conacher (University of Western Australia); Bob Fagan (Macquarie University); Susanne Freidberg (Dartmouth); Bill Friedland (University of California, Santa Cruz); Harriett Friedmann (University of Toronto); Joerg Gertel (University of Leipzig); David Gibbs (University of Hull); Barbara Harriss-White (University of Oxford); Alex Hughes (University of Newcastle); Lucy Jarosz (University of Washington); Geoff Lawrence (University of Queensland); Becky Mansfield (Ohio State University); Warren Moran (The University of Auckland); Mike Morris (University of Cape Town); Dominic Power (Uppsala University); Graeme Wynn (University of British Columbia).

Chapter 1

Introduction: Mapping the Concept of Globalising Networks

Christina Stringer and Richard Le Heron

Although we often hear a great deal about globalising networks in the current economic geography and agri-food literatures, the intellectual work of globalising agri-food networks seems to have somehow slipped below the horizon. This book explores the globalising connectivities implied in the notion of globalising networks with a view to drawing out some of the complexities and contradictions evident when the conception is harnessed with that of commodity chain. By opening up space to engage with the constitutive processes that produce networks that might be labelled globalising the collection prioritises an examination of what it has meant to produce food and fibre in emerging capitalist societies. Instead of considering globalising networks as an exclusively contemporary phenomenon, the collection illustrates from research into both the past and present agri-food commodity chains some intriguingly similar dimensions in the constitution of responses to the challenges of securing value circulation and exchange across great and short distances.

We approach the idea of globalising networks first by briefly outlining a genealogy of commodity chains and networking. By the early 1980s there was much debate internationally about uneven development and uneven exchange, a great deal springing from the structural positioning work of the world systems writers who divided the world into a capitalist core and periphery, with ambiguous cases assigned to the semi-periphery. Despite the inadequacies of the functionalist imaginary that underpinned this schema, it does highlight a key world-level insight, that conditions of production and consumption are uneven. In this literature the paucity of diagrams summarising capital circulation perhaps constrained the imagination of researchers. Whatever the reason the notion of focusing on commodity circuits had little appeal and was for the most part only dimly perceived. In this context it is hardly surprising that a conceptual and methodological strategy – the depiction of the chain of links in commodity circulation – should be enunciated. Importantly it involved a metaphor that was visual and which readily revealed connectivity. Whether the genesis of the idea emerged from the Marxist political economy of agriculture or the world systems communities the intention was, as Wallerstein (2000, 221) remarks, 'to underlie a basic process of capitalism: that is it involved *linked* (emphasis added) production processes that had always crossed multiple frontiers and that had always contained within them multiple modes of controlling labour'. It was not until the mid-nineties, however, that agri-food commodity chain analyses had been mainstreamed.

The other trajectory of interest is that of the network. This idea, surfacing prominently in the 1990s in industrial and broader economic geography, had appeal to those revisiting the grounds for competitiveness, regionally, nationally and internationally. The apparent successes of new industrial districts in California and North Italy for example attracted scholarly scrutiny in the late 1980s, with the place of networks, a concept soon to be positioned with ideas such as social capital, trust, cooperation and so on, deriving much of its attraction from initial attachment to proximity rather than connection. Again, the infusion of networks and networking concepts into agri-food circles proceeded slowly.

A new generation of researchers had begun to take seriously some of the challenges presented by post-structural, feminist and sociology of science writers. Agri-food research practice began to unpack the very categories upon which both the commodity chain and network ideas rested. The first conceptual point we make is that a succession of important steps of a theoretical kind began to emerge. These steps include: a widening of the scope of networks and networking through visioning both additional horizontal and vertical dimensions of networks; the variety of theoretical approaches that could be brought to bear upon understandings of commodity chains and networks and networking was added to; attempts were made to integrate alternative approaches in order to better comprehend the contours of use and exchange value production and value circulation; different value propositions were used as the point of entry into analyses and investigation; and, the spatiality of investigation was extended with increasingly sophisticated studies of geographic scaling and politics.

In the foregoing discussion we deliberately chose not to problematise the notion of commodity chain. Here, as our second conceptual intervention, we recognise the plurality of conceptualisations available. In the academic literature the earlier concept, commodity chain, has been matched by an astonishing rollout of two parallel concepts, one very instrumental in character and which is probably very widely understood, that of supply chain, and the other, value chain, which may seem to promise a lot, but is in fact under used and poorly understood outside purely business circles. The commonality of scope, in so far as each is an attempt from different starting positions to explore different though often overlapping dimensions and concerns (from usually sharply divergent political and ethical positions) has led to much confusion over what is being discussed, and worse, frequent conflation of the terms when using only one. A Google concept search shows that the commodity chain idea *per se* has little traction, it is an academic notion favoured by a small community of scholars. Next and considerably more widely used is the value chain notion, which while integral to business models, is clearly because of its infancy still limited in application to within the business community. By far the most extensively developed framework is that of the supply chain; anchored in physical flows, responsive to logistical effort, amenable to governance on sophisticated levels using a variety of at-a-distance systems, it has become a distinctive unit for organising competition especially in the international arena. We distinguish amongst the three positions because although the contributors overwhelmingly adopt some sort of commodity chain approach, several set their research in the realm of the value chain and supply chain literatures.

The overarching emphasis of commodity chain analysis, as derived from the traditions mentioned, is to enquire into the conditions under which value is created, realised and reinvested in an increasingly interconnected world, with the intention of revealing the implications for labour and bio-physical processes in different contexts. This is in effect a social or political economy view, concerned mainly with establishing how actors are involved in a commodity's complex movements, evaluations and valuations, with regard to intersecting ecologies. In contrast, the central proposition of the value chain school of thought is centred more on the performance of individual actors. The interest is in finding ways to generate and appropriate, for the actor, value that is circulating in the commodity and cultural circuits. The rise of supply chain management as a specific area of business research, usually without explicit consideration of power relations amongst business entities, indicates the utility of the concept for business and for interrelated businesses. It is around points such as these that divergence with commodity chain thinking begins.

The third dimension to our conceptual mapping is the manner in which this book and the authors deploy globalising networks in research on agri-food commodity chains. They do this in four distinctive ways and thus *Agri-Food Commodity Chains and Globalising Networks* contributes to contemporary frontiers in agri-food research. First, the authors throughout the book acknowledge the contributions of Australasian agri-food researchers to the study of the circulation of value and highlight the networks of production and consumption, trade and finance and their interrelations and interactions. This reflects the strong tradition of new political economy within agri-food research. But it is also a methodological statement, in that to name globalising networks in this manner is potentially constitutive of understandings that shed light on capitalist accumulation processes. Second, the authors invoke an organisational focus in order to explore the development of the metrics of calculation and institutional embedding that underpin the rise and functionality of governance technologies directed at assisting value circulation. This leads into dimensions of competition under new rules of economic and cultural standardisation in production and consumption. Third, while the place of regional networking in creating conditions for possible agri-food actor participation in local provisioning or far distant supply had received attention, empirical evidence about such developments is still sparse, undoubtedly affected by the limited pool of researchers and funding. Fourth, the inexplicable bias against historically grounded research in agri-food research, and in economic geography more generally, has led to only a handful of studies informed by theoretical developments on commodity and cultural circuits and still fewer taking up the challenge of understanding the changes in networks and practices of networking. That this prevents meaningful dialogue over the emergence of long run trajectories around agri-food activities is an understatement. *Agri-Food Commodity Chains and Globalising Networks* makes specific contributions to these four contemporary frontiers, drawing on new understandings that have come from a number of longstanding and major research programmes undertaken in a range of different countries. The authors' contributions reflect their distinct regional perspective as they explore new initiatives and responses to developments, whether internally or externally driven, within agri-food chains and globalising networks. The book is organised into four sections:

Section 1: Globalising Networks and Agri-Food Chains

This section consists of three chapters that provide lens upon the formation of value through globalising networks. In particular, the authors examine ways in which the case study actors are responding to globalisation pressures and what this might mean for actors in developing countries seeking to participate more fully in globalising networks.

In Chapter 2, Charles Mather contends that the global value chain concept of 'upgrading' has particular relevance for developing countries. He argues that the ability to upgrade arguably determines whether firms, regions and countries are successfully inserted into the globalising economy. The chapter focuses on South Africa's squid industry, which has since the late 1980s established itself as a reliable supplier of high value fish products to Spanish and Italian markets. The chapter describes the process of upgrading squid production from the late 1980s and concludes that in contrast to early upgrading initiatives more recent initiatives have not lead to significant real benefits.

From a very different starting point, Andreas Stamm explores the way the Millennium Summit of 2000 commits the international community to a strategy of accelerated poverty reduction. One of the promising sectors within a strategy of pro-poor growth is agribusiness as it allows developing countries to capitalise on their natural production advantages and through its backward linkages directly affects the lives of the rural poor. Based on empirical research in Sri Lanka and taking the value chain approach as conceptual framework the chapter examines, under which conditions agribusiness based growth actually leads to an accelerated reduction of poverty. While the evidence assembled in the chapter is still quite general, the importance of originating networking and related power asymmetries for farmers in Sri Lanka remains clear.

The shift to fewer and larger upstream suppliers has been a prominent theme in many recent analyses of value chain restructuring, especially in agri-food sectors. In developing countries, these tendencies are usually played out in the context of smallholder exclusion, with a corresponding shift of production to larger enterprises. However, in some instances powerful counter-currents can be observed. In the South Indian tea plantation sector, larger plantation enterprises have faced crisis due to low world prices, rigid management structures and high costs of production. In this environment, small tea growers appear to hold a substantial competitive advantage vis-à-vis larger estates, resulting in a new phase of restructuring based around innovative alignments between centralised processing units and smallholders. The chapter addresses ongoing debates on the relative efficiencies of localised modes of agricultural production within a changing global regulatory framework. Reviewing these issues in their broader terms, Neilson and Pritchard contend that the South Indian experience emphasises the importance of thinking about value chain issues in terms of the institutional histories of how particular production complexes have evolved in specific regional economies. It highlights the role of geography in creating mosaics of difference in the ways production systems are structured.

Section 2: Governance and Globalising Networks

The chapters that make up the second section of the book examine developments within globalising agri-food networks from a governance perspective. The authors examine different regulatory and governance structures ranging from the emergence of calculative practices to forms of industry led self-regulation and development.

In Chapter 5, Christopher Rosin approaches the yerba mate (*Ilex paraguariensis*) commodity chain in Brazil and Paraguay within the MERCOSUR customs union from a convention theory perspective. Undertaking a series of interviews with farmers, he assesses the role of producers and their capacity to influence the emerging conventions of exchange. Factors such as relative economic and social power and existing market conditions are limitations on producer influence. Despite such constraints, producers are shown to affect the development of conventions through their willingness to produce yerba mate or to adopt alternative crops. The analysis demonstrates the insights of convention theory to the emergent nature of commodity chains.

Chapter 6 examines the emergence of audit systems certifying 'good agricultural practice' within the global agri-food system. These audits ensure consumers of food produced in a sustainable (environmentally and socially) manner without reference to organic practices. Using data from 36 New Zealand kiwifruit orchards, Christopher Rosin, Hugh Campbell and Lesley Hunt provide a detailed examination of the impact of the Eurep-GAP audit on farm management. As much as good practice, orchardists associate the audit with increased paperwork – an irritation affecting relations with other commodity chain participants. They conclude that, while Eurep-GAP can promote improved sustainability, it also impacts the power relations of New Zealand's agriculture.

An example of industry led self-regulation is illustrated in Chapter 7. Paula Blackett and Richard Le Heron discuss the formulation and implementation of an industry regulated Accord in 2003. The Accord was designed to facilitate improved management of the environmental externalities of farming practices on New Zealand dairy farms. The key rationale behind the Clean Streams Accord was to afford the 'clean green' image of New Zealand dairy farming and food production some protection in the market place. The authors explore some of the key drivers behind the Accord and comment on the current institutional arrangements and challenges around managing environmental impacts of dairy farming practices and in particular dairy shed effluent discharges to waterways.

Eugene Rees, in Chapter 8, explores how New Zealand seafood enterprises have evolved as they simultaneously negotiate local regulations that constrain resource extraction as well as the globalising world economy. Conventional arguments for the institution of quota management and transferable quota propose that re-regulation leads to efficiency, and by implication, profitability gains. While this literature is replete with discussions concerning the nature of the efficiency gains, there is little discussion concerning the form that fishing enterprises will take or how they will go about instituting efficiencies and profitability gains. Moreover, changes in fisheries and aqua-commodity production which have been *made possible* by a quota management system have passed largely unheralded. The discussion moves beyond equity debates to examine how firms have gone about securing growth when operating in the New Zealand fishing sector by reconfiguring their activities throughout an increasingly global commodity chain.

In Chapter 9, Nicolas Lewis uses the case of the wine industry in New Zealand to examine the properties of 'industry' as an economic space of governance – a space that has meaning as an identity, a spatial imaginary, and a subject as well as an object of calculation and governance. He links arguments about the significance of industry scale governance to notions of industry as identity and neo-liberal rationalities of governing at a distance. Furthermore, he challenges economic geographers to recognise industry as more than an essential economic category, and to recognise 'spaces of government' such as industry as an alternative language to 'scale' for identifying meaningful social collectivities for analysing contemporary political economy.

Section 3: New Agri-Food Configurations from Networking in Regional Contexts

This section brings together researchers who examine levels of embeddedness between local and regional networks. They examine ways in which growers and farmers can, for example, overcome challenges through the establishment of learning networks and clusters in order to facilitate innovation and differentiation.

Despite continued interest in global commodity chains, the globalisation of agro-food systems is uneven and mediated by local relationships. In Chapter 10, Brian Ilbery and Damian Maye examine the development of 'local' foods in the 'lagging' region of the Scottish-English borders. Using institutional food maps, an important boundary effect is identified, with different sets of institutional processes operating either side of this international border. However, such a boundary effect is not apparent among specialist food businesses. The empirical findings thus suggest that there is a mismatch between the activities of food-related institutions in the Scottish-English borders and other key players in the local food economy.

In Chapter 11, the regional focus shifts to the Pyrzyce (West-Poland) and Elbe-Elster (Eastern Germany) rural districts. In this chapter, Elmar Kulke illustrates that both regions posses a regional agrarian cluster which reaches beyond the local scale and substantially affects the regional economy. The clusters are based on material connections and immaterial co-operations and information transfer. Remarkably, in these immaterial flows, local forums are of great importance. The results illustrate that the proximity of the actors alone is not the determining factor for immaterial networks and that supra-regional linkages play a key role. It is evident that not every farm which has material connections in a region also possesses immaterial networks. The cluster analysis shows that not only large traditional corporations but also younger individual enterprises with considerable immaterial linkages can achieve competitiveness and good power positions. Ideally the best practice for farms is therefore to be embedded in local and regional, as well as in supra-regional immaterial networks, which are based on equality and reciprocity.

In Chapter 12, Mairi Jay examines how dairy farmers in New Zealand are responding to competing pressures around agricultural sustainability and what type of new configuration might emerge. Farmers in New Zealand face pressure to maximise low-cost forms of production in the face of global economic competition.

At the same time, they experience domestic pressure for improved environmental performance and animal welfare. The twin pressures of environmental improvement and economic efficiency have resulted in varying responses by farmers. The dairy industry has encouraged a model of environmental efficiency in accord with business rationality whereby farmer milk suppliers undertake measures that reduce pollution of waterways. An alternative approach views the environmental aspects of sustainable agriculture more broadly by posing the question: 'Can what is being done now, still be working successfully in 100 years?' While the two approaches differ in starting point and philosophical thrust, Mairi Jay explores the extent to which they differ in practice and suggests implications for ecological modernisation theory.

Chapter 13 examines how independent winegrowers and their wine regions are competing against the economies of scale and scope achieved in large consolidated firms. Jerry Patchell discusses what types of markets, institutions and economies of scale need to be constructed, taking into account the need of winegrowers to balance collective reputation with individual differentiation. Evidence is drawn from the wine regions of Bordeaux, Napa, and Chianti Classico, discussing their successes, internal problematics, and requirements for external governance and assistance.

Neil Reid and Michael Carroll examine the development of an industrial cluster focused on the greenhouse industry in northwest Ohio in Chapter 14. The cluster was the idea of university researchers and was designed to help the industry address some significant competitive threats, including intense international competition and escalating energy costs. However, from very early on in the evolution of the cluster it became apparent that getting a disparate group of growers to think and act collectively was going to be a major challenge. The industry was imbued with low levels of social capital that threatened efforts to organise the industry as a functioning cluster.

Section 4: Globalising Networks and Agri-Food Chain Development and Innovation

The final section of four chapters explores development and innovation in agri-food chains from New Zealand – a resource periphery – perspective. Chapter 15 is a contemporary examination of how, Fonterra, New Zealand's largest company has overcome its peripheral location. In contrast, Chapters 16, 17 and 18 take a historical perspective to illustrate New Zealand's incorporation into shipbuilding and grass seed commodity chains during the late nineteenth and early twentieth centuries.

One of the challenges facing companies in resource periphery locations is overcoming the disadvantages of distance. Chapter 15 explores the way in which Fonterra Co-operative Ltd, New Zealand's largest company, has moved beyond geographical location 'on the edge' to become a leader in the global dairy industry. In particular, the chapter examines how the company has expanded internationally through a series of innovative partnerships with international dairy companies.

One of the earlier statements about commodity chains emerged in the 1980s out of Wallerstein's world systems research project. In Chapter 16, Michael Roche explores successive refinements to global commodity chain concepts from within the world system school. He focuses more closely at the shipbuilding commodity chain, in particular at the way in which New Zealand kauri forests were incorporated into the commodity chain in the 1830s. Roche concludes by examining how the New Zealand segment of the shipbuilding is used as a basis for some general observations about global commodity chains.

Vaughan Wood in Chapter 17 examines a distinctive agri-commodity, namely grass seed, with a focus on the export from New Zealand of cocksfoot (orchard grass) seed, which was highly desirable to British grassland improvers during the years 1880–1930. The emergence of an export market for 'Akaroa' cocksfoot led to consolidation of the supply chain by New Zealand company players and the tightening of quality controls, but ultimately environmental constraints on the production process caused this formerly sought after 'brand' to be displaced by cheaper international competition.

The new international division of labour in the nineteenth century globalising world was driven by the emergence of commodity chains that generated 'improvement' in the lands of the imperial periphery as these were drawn into trade in grass-based products for the metropole. Chapter 18 uses the concepts of centres of calculation and brand value to explore a little known aspect of the imperial web: the increasing exchange of the grass seed upon which new commodity chains depended. Eric Pawson explores how value was added in this arena, using the New Zealand/ Britain link between 1850 and 1930.

Acknowledgments

During the course of the book's preparation we have received much support from a number of quarters. We are especially grateful to Igor Drecki for finalising the cartography work for the book. We are also very grateful for the secretarial support of Susan Sum and Beryl Jack throughout the preparation of the book. Colleagues in the Department of Management and International Business and the School of Geography, Geology, and Environmental Science at The University of Auckland provided helpful advice and suggestions at various stages.

Reference

Wallerstein, I. (2000), *The Essential Wallerstein*, New Press, New York.

Chapter 2

Upgrading in South Africa's Squid Value Chain

Charles Mather

Introduction

The process of *upgrading* is central to the global value chain framework. In its earliest formulation, upgrading described the process of firms or sectors taking on more complex and profitable production functions within a value chain. Gereffi's (1999) work on the clothing industry represented an early blueprint for upgrading; in his research he was able to show how firms initially involved in simple assembly were able to move 'up the chain' into the production of entire garments and finally into the design and sale of branded merchandise. For clothing manufacturers each phase of upgrading corresponds to a higher level of value-adding.

Research on global value chains since the late 1990s has both extended and refined the concept of upgrading. The work of Rammohan and Sundaresan (2003) has stressed the need to link upgrading to broader social and development outcomes (see also Bair 2005; Bair and Gereffi 2003; Humphrey *et al.* 2004; Harilal *et al.* 2006). For Rammohan and Sundaresan (2003, 906) this may be achieved by 'socially embedding' value chains in a way that goes beyond an economic analysis to a consideration of the impact of upgrading on 'the lives of peripheral workers'. Nadvi's review of several Asian and African agri-food, clothing and textile value chains also addresses the question of whether participation in global value chains – which often requires upgrading – can 'deliver pro-poor outcomes – not only in terms of income gains, but also with respect to employment stability, income security and working conditions' (Nadvi 2004, 21). The results of his survey suggest that insertion into the global economy has seen the creation of new jobs and income gains for workers. Yet there is also evidence to suggest that the benefits to workers may depend on where you are in the chain. In addition, all workers are vulnerable to the process of casualisation.

A second direction of scholarship on upgrading has refined the concept and also identified the conditions under which upgrading is more or less likely to occur. Researchers working in this direction have suggested that it is useful to identify four types of upgrading: process, product, functional and inter-sectoral upgrading. This new approach to value chains has also been associated with a more complex formulation of chain governance that goes beyond the initial categories of producer and buyer driven chains. Gereffi *et al.* (2005) have suggested a more 'complete typology' of chain governance that captures the complex relationships between buyers and producers based on the complexity of the transaction, the ability to codify transactions and capabilities in the supply base. The significance of these new categories – market, modular, relational, captive and hierarchical – is that the possibility of upgrading is better under certain governance regimes than it is under others (Humphrey and Schmitz 2001). This new framework for both upgrading and chain governance has generated considerable comment and critique (e.g. Gibbon and Ponte 2005; Bernstein and Campling 2006).

This chapter hopes to contribute to the debate on upgrading through an analysis of recent changes in South Africa's squid export sector. Before the mid-1980s the industry was focused on supplying squid (*chokka*) as bait to other fishing enterprises based on a very low level of technology and poor returns. When Italian fish buyers recognised the potential of the local squid species as an export commodity, the industry was rapidly transformed to meet their demands for a freshly frozen product. More recently, upgrading has taken the form of responding to new process requirements by the European Union associated with food safety. The squid case raises questions about the usefulness of the new categories of upgrading and the possible link between upgrading and livelihoods. Rather than trying to fit the squid case into an upgrading category, the chapter focuses on changes in the impact and pattern of upgrading over time. Initially, the squid sector was extraordinarily successful in upgrading from a very low level of technology and quality to a sophisticated sector producing a high value product for export. This phase of upgrading had a significant impact on the income generated by the sector and created new employment opportunities for fishermen. The upgrading that is taking place at present, however, is not leading to income gains and has instead become a requirement for *participation* in this global value chain. Local exporters are pursuing other options for upgrading but these are limited by the nature of the commodity and by European buyers. With regard to the development impact, the case focuses on working conditions for squid fishermen. A consequence of upgrading was the creation of several thousand new jobs on squid boats, yet working conditions are poor. Improving working conditions is more likely to occur through local regulatory efforts rather than through a process linked to upgrading or participation in this value chain.

The first section of the chapter describes the global value chain for squid. This section emphasises changes in the geography of squid production and consumption and South Africa's position as a small, but nonetheless important player in the global squid trade. In section two, the process of upgrading in the sector is described. In the final section, the emphasis shifts to a discussion of more recent efforts and obstacles to upgrading. This last section also considers the impact of upgrading on squid fishermen. The research is based on secondary material on the squid sector and interviews with local exporters and squid boat owners in Port Elizabeth, St Francis Bay and Cape Town.

Squid Value Chain

There have been important geographical changes in the production and trade of squid products in the last 30 years. Up to the mid-1980s, squid production was dominated by Japan: in 1980 this country's fleet was responsible for 60 percent of global squid production (Globefish 2003). A large proportion of the squid captured by Japanese trawlers occurred in waters where restrictions on foreign fleets were not in place. From the mid-1980s, however, the situation changed and while Japan remains an important producer of squid, it has lost considerable ground to new competitors including the Republic of Korea, China, Argentina and Peru. The key reason for the decline in Japanese squid production is related to the declaration of 200 miles Exclusive Economic Zones (EEZ) by many developing countries, which effectively restricted foreign fleets from access to fish resources. Japan's share in squid production declined from over 60 percent in 1980 to just over 18 percent in 2002 (Globefish 2004).

The global trade of squid reveals a different picture. Of the top three producers in the world, Korea is the only country involved in the sale of squid to other markets. The other major producing countries, notably China and Japan, consume all or most of the squid captured by their fleet. The largest exporters behind Korea are Spain, Argentina, Viet Nam, Thailand and Morocco. The main areas of squid consumption are south east Asia and Europe. Spain, Italy and the United States are the largest importers of squid products in value terms. These three countries imported almost 69 percent of the total squid traded in the period 1999–2003 (Globefish 2004). The Republic of Korea, France, Greece and Portugal follow some way behind. Regionally, European countries combined are the largest importers, purchasing an average of 74 percent of globally traded squid products over the period 1999–2003 (Globefish 2004).

South Africa is a small player when it comes to overall production of squid. Globally the local industry is only twentieth on the list of squid producers by volume. Yet many of the largest producers consume what they catch and are not active in the global squid trade. In contrast, South Africa's production is all exported and no locally captured squid is consumed in the domestic market. As a consequence, South Africa's position in the global trade of squid products is considerably more important. In Spain and Italy, the main markets for local exporters, South Africa is the third and fifth largest supplier respectively (Globefish 2003).

The total production of squid in South Africa varies considerably depending on the state of the biomass, which appears to be shaped by climatic factors (Roel and Butterworth 2000). Average production in the period between 1989 and 2003 was close to eight thousand tons. Both volume and value fluctuate considerably, but on an upward trend. Average yearly volumes increased by 24.5 percent between 1993–98 and by the same amount between 1999–2003 (Figure 2.1). The value of exports increased by 28 percent for the same period with 2003 being an exceptionally good year for the industry with earnings of over R430 million (Oceanafrica 2005b). The price of South African squid on overseas markets is determined by local production volumes and the amount of squid being supplied by other countries. Although there are small seasonal and annual changes in the price of squid in Italian and Spanish markets, these have remained very stable over the last five years. The Rand – Euro exchange rate has a far greater influence on exporters' earnings.

In assessing the global significance of South Africa as a squid producer, the issue of quality and squid species needs to be added to the equation. There are two key

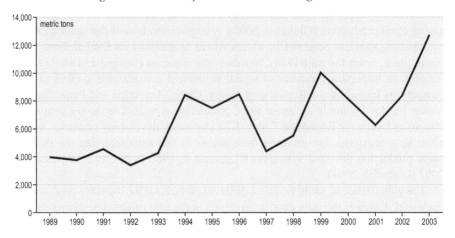

Figure 2.1   Volume of squid exported from South Africa, 1989–2003

Source: Globefish 2004.

squid species, *Loligo* and *Illex*. The species caught in South African waters is *Loligo*. While both varieties are imported into Europe in large volumes, the demand for the *Loligo* species is far higher, which is reflected in a significant price difference between these two species (Table 2.1). Italy has always favoured *Loligo* squid and demand in the Spanish market is increasingly shifting towards this species. The demand for *Loligo* in Italy and Spain has allowed South African exporters to increase supplies of frozen squid to both countries (Figure 2.2).

Table 2.1   Spanish squid imports: country source, volume and price

Country of origin	Squid species	Metric tons		Euro/kg	
		2003	2004	2003	2004
Morocco	Loligo	1,976	736	6.83	8.82
South Africa	Loligo	3,095	5,035	4.63	4.56
India	Illex	8,435	8,849	2.16	1.61
China	Illex	2,737	3,171	1.52	1.82
Korea Republic	Illex	1,614	1,649	2.54	2.22
Falklands/Malvinas	Illex	19,743	12,838	2.01	2.36
Peru	Illex	3,346	3,272	1.27	1.47

Source: Globefish 2004.

The structure of the South African squid export chain is relatively straightforward. Around 130 chokka boats harvest squid for 11 months of the year and the fish is sold to locally based processor-exporters, all of which own large cold storage facilities for the frozen product. The number of local processor-exporters involved in the export of squid is relatively small and there is considerable competition for fish. As a result,

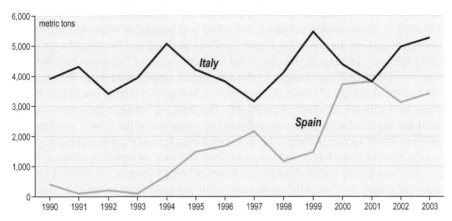

Figure 2.2   South African squid exports to Spain and Italy

Source: Globefish 2003.

most of these companies are also directly involved in squid production through their own vessels. In Spain and Italy the frozen squid blocks are sold to independent buyers who then sell on to the fresh produce markets or directly to restaurants and catering companies. Despite the absence of large retailers, the South African squid value chain is 'buyer driven'. Italian and Spanish buyers set the price for squid and they are responsible for setting and monitoring quality standards. Local exporters also suggested that these buyers were better organised and had in depth knowledge about the season's production volumes, which they then used to negotiate prices.

South Africa's Squid Sector: Origins and Upgrading

South Africa's squid export sector is young relative to the country's broader commercial fishing sector. Local fishermen have always harvested squid, but almost exclusively as bait for other line fishing activities. In the 1970s, Japanese and Taiwanese trawler vessels recognised the value of the resource and exploited it as a food commodity for their own domestic markets (Sauer *et al.* 2003). Japanese vessels were also involved in the first assessment of the squid biomass, which they estimated to be approximately 35,000 tons. By the late 1970s foreign trawlers were restricted from South African waters when the country declared a 200 mile Exclusive Fishing Zone, in line with international treaties on the protection of national fishing resources. The exploitation of the resource for human consumption by South African fishermen followed shortly afterwards but took off in 1985 and 1986 with a decline in the value of the rand and the simultaneous discovery of a market for South African squid ('*calamari*') in Italy (Oceanafrica 2005a).

When the squid sector was first integrated into global markets in the mid-1980s the barriers to entry were very low. Squid was caught using hand lines (jigging) and ski-boats or deck boats, some of which in turn supplied larger 'ice vessels' at sea. In parts of the Eastern Cape ski-boats landed their squid harvest onto the beach and then supplied the fish in crates to shore based factories in Humansdorp and Port Elizabeth (Figure 2.3). Once landed on shore, the squid was packed in trays and frozen for export. Processing the squid for export involved packing and freezing the fish into 10–12 kilogram rectangular blocks. From the perspective of European buyers there were considerable disadvantages to this method of squid production. When squid was harvested using a ski-boat, it often spent considerable time at sea before being frozen into a 'block of squid'. Remarkably, this time lag between the capture of the squid and the freezing process was reflected in the colour of the 'block' produced in the shore-based factory. A long delay between the capture of the fish and the freezing process resulted in a white or light brown coloured 'block' of squid. In contrast, when the squid was frozen within three to four hours of being captured, the frozen block of squid had a dark or 'chocolate' appearance. This unique feature of squid physiology meant that a dark coloured block of frozen squid indicated a 'freshly frozen' fish. For European importers a uniform dark chocolate colour – or 'CJ' grade according to the industry's standards – was therefore a proxy for both process and quality. When the colour of the squid had shades of both brown and white the block was downgraded to a 'C' grade, which carried a price penalty. Squid that was completely white, indicating a long delay between capture and freezing, carried the 'Numbers' grade and was normally not suitable as an export commodity.

In response to the demands of European buyers, several squid fishermen purchased freezer boats or they upgraded existing deck boats with blast freezers and cold storage facilities. The first fishermen to catch squid using boats with blast freezers were extremely successful in achieving the 'CJ' grade for up to 95 percent

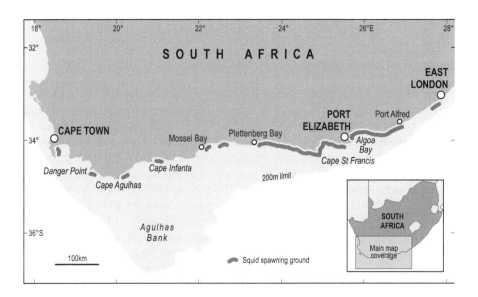

Figure 2.3   Squid spawning grounds, South Africa

Source: School of Geography, Archaeology and Environment Science, University of Witwatersrand.

of the squid captured. These early 'pioneers' were richly rewarded for the new innovation in squid capture and processing: several remembered being paid up to seven times more for sea frozen squid than they were for fish frozen in shore based factories (interviews). The large difference in price between sea frozen squid and 'wet fish' – the term used for squid processed in shore based factories – was a key driver in the transformation of the rest of the fleet from one based on ski-boats and deck boats to one based on more expensive and sophisticated freezer vessels. In the late 1980s, ski-boats and deck boats dominated the industry; by 1998, 60 percent of the fleet consisted of freezer boats with deck boats and ski-boats making up the remaining 40 percent of the fleet. In 2004 there were no deck boats fishing for squid and only a very small number of ski-boats (Figure 2.4).

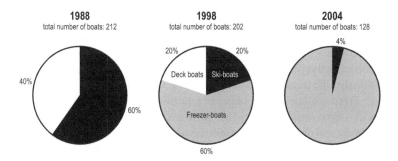

Figure 2.4   Changing technology for squid harvesting, 1998–2004

Source: Oceanafrica 2005b.

Italian buyers played a key role in driving the upgrading process in this sector of South Africa's fishery. Local exporters played their role by passing higher prices down to boat owners and by assisting them to upgrade either through advice or through loans. As has been reported in other commodities that experience upgrading, those producers unable to upgrade were increasingly excluded from the squid value chain (cf. Gibbon 2001). Most squid boat owners faced the choice of upgrading or leaving the industry altogether. A key problem was that the traditional method of capturing fish failed to deliver a consistent chocolate colour. Indeed, boat owners using traditional methods found it increasingly difficult to convince exporters to take their product. As one boat owner recalled, keeping the fish on ice 'made it look terrible – the skin used to break and it would go white and it then wouldn't colour up; the quality was uneven and they (exporters) wouldn't take what you caught' (interviews).

While the demand by exporters for sea frozen squid was the primary driver of this transformation, there were additional advantages for boat owners who upgraded to factory vessels. The introduction of freezer boats allowed fishermen to catch squid year around and in deeper waters, which was not possible using a ski-boat. Deep-water fishing for squid requires the use of an additional innovation in the form of a 'parachute', which acts as an anchor. Smaller craft are, in contrast, restricted to shallower waters where squid are present, but only during 2 or 3 months of the year. The result is that while freezer boats can fish for most of the year, ski-boat owners can only be economically active during a relatively short period of time.

A second advantage of using freezer boats to catch squid is that it allows boat owners to exercise almost complete control over the production process. When fishermen catch sufficient fish for a 10–12 kilogram block, the squid is carefully packed and placed in the blast freezer. After the fish is frozen it is removed from the blast freezer and placed in cold storage holds on the ship. Once the squid is offloaded at a port it only requires fresh water glazing and packaging before it can be exported. Ski-boat owners, in contrast, were – and the few remaining producers still are – dependent on supplying shore based freezer facilities with a fresh and perishable fish product that often fails to meet export quality standards. Finally, technological developments associated with freezer boats have acted as a mechanism for excluding ski-boat fishermen. Freezer boats use powerful lights to attract the squid at night. Although lights can also be attached to ski-boats, these are much less powerful. According to one of the few remaining ski boat owners, the powerful lights on freezer boats effectively attract the squid 'right from under my boat' (interviews).

Traditionally, the transformation of the squid sector is described in terms of 'fleet evolution'. From the late 1980s, according to this account, the industry shifted its market focus – from squid as bait to high value export commodity – and in technology – from ski-boats and deck boats to sophisticated freezer vessels with high-technology fish finding equipment. In these accounts this sea change is described as the natural evolution of a fleet from one based on low technology and inefficiencies to a modern fishery. Yet the rapid change in the technology base for squid fishing is far more significant than a process of 'fleet evolution'. From the late 1980s squid harvesting became a specialised sector of South Africa's fishing economy with purpose built boats and an infrastructure aimed at meeting the demand for a specific quality of fish product. This process was *driven* by the demand of foreign buyers for 'chocolate' squid and is a good example of upgrading and global integration.

On Upgrading and Development in South Africa's Squid Sector

The upgrading process that occurred from the mid-1980s, and which was driven by European buyers, is qualitatively different from the current demands being made on squid producers by the European Union. While Italian and Spanish importers assess the quality of the squid through a visual inspection, the new process standards being imposed by the European Union are in line with how fruit and other fresh product value chains are being monitored and assessed. These methods of monitoring are both hands-off and more 'objective' than the quality requirements imposed by European buyers.

In its efforts to ensure that food products reaching the region are safe for human consumption, the European Union has accredited the South African Bureau of Standards (SABS) to test all local squid catches destined for Europe. SABS agents take a sample of each landing and test it in their laboratories for any contaminants. If the sample inspected by the SABS contains impurities of any kind the fishermen risks having the entire consignment destroyed. Traceability, which allows all agents along the value chain to trace quality problems back to the boat, is now built into the production process. Before the squid is blast frozen, a tag is placed within the squid block indicating the date of freezing as well as the ship's identification number, which must registered with the European Union. The identification of each block of squid allows exporters and importers to address quality problems by removing fish found to be contaminated or unfit for human consumption. Exporters control quality problems through financial penalties. If the quality is poor or if the importer rejects the consignment, the financial losses are passed back to the boat owner. With a strong local currency, the emphasis on minimising quality problems has intensified and some exporters may encourage the skipper to be trained on how to achieve good quality squid and how to avoid hygiene problems through Hazard Analysis and Critical Control Point (HACCP) training. Squid producers confirm that new standards and process requirements are continually being introduced. For instance, boat owners are now required to install monitoring devices in their freezer holds to ensure that the temperature does not exceed a certain threshold. This information is then 'attached' to the fish consignment as a record of the production process.

The process requirements now required of squid boat owners and exporters have much in common with what other agri-food exporters are facing in global value chains. Unlike the previous phase of upgrading, these new requirements are assessed objectively, often through third party auditors and involve huge amounts of paper work and, increasingly, digital information on the environmental conditions in which the product has been stored. This second phase of upgrading is also different from the first in that it has not generated income gains for boat owners or exporters; it is instead a criteria for chain participation. The costs associated with this phase of upgrading must be borne by boat owners and exporters.

The scope for upgrading that has tangible outcomes in terms of income has become very important in the last two years for reason that are associated with dramatic fluctuations in the local currency. In the period between the 2001 and the middle of 2002 the South African currency weakened significantly against the dollar and other major currencies leading to a boom for the local squid sector. By the middle of 2002, the South African currency had weakened to a level whereby squid prices in Rand terms were above R50 a kilogram. In the period since 2002 the South African Rand has strengthened considerably in against the dollar, which has led to far lower prices for the local export sector. Indeed, the price of South African squid has returned to the levels recorded before 2000 (Figure 2.5).

Local exporters have responded by attempting to upgrade in ways that will improve their export prices in the face of a stronger local currency. They face a daunting task: the price of South African squid has remained very stable over the last 7 years ranging between US$4.25 and US$5.00 despite fairly dramatic changes in the overall stock of squid on European markets. In other words, it appears that stock levels have only a limited impact on squid prices in Spain and Italy. Several exporters have focused on improving the quality of squid produced. Even though the standards are already very high, one exporter has produced a manual, which is used to train crew on how to pack squid for optimal results in terms of quality. Boat owners and exporters have also encouraged fishermen to attend HACCP courses

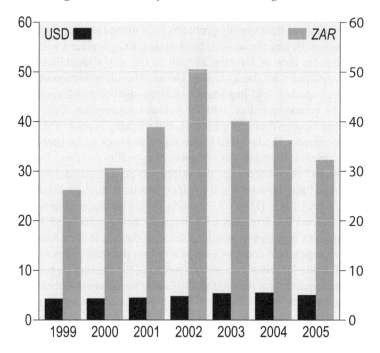

Figure 2.5   South African squid exports by price and volume

Source: Compiled from Globefish 2005b and unpublished industry sources.

to prevent any rejections from either the local EU appointed agency or overseas exporters. Several exporters have also embarked on a branding campaign in an effort to create consumer loyalty and perhaps eventually increase export prices. Other efforts to upgrade include producing smaller blocks of squid, which are more convenient to restaurant owners and caterers who deal with smaller volumes of squid. Finally, several exporters have begun experimenting with individually packed squid for the Japanese market, although there are technical difficulties in producing an individually packed, chocolate coloured squid.

It seems unlikely that these efforts will make a marked difference to the export earnings of the squid sector. Efforts to improve the quality of squid production are limited because the scope for improvement is very small. Exporters already expect 95 percent of the squid to be 'CJ' or 'chocolate' grade, which means that any gains in quality will only have a small impact on the returns of exporters and boat owners. The idea of branding the 10–12kg squid blocks depends on the cooperation of Spanish and Italian buyers. Securing the cooperation of European buyers may be difficult because it is not necessarily in their interests to sell a branded squid product. The success of European buyers depends on the consumer loyalty of caterers and restaurant owners, which is based on *their* reputation as buyers of high quality fish, and not on their ability to source a branded product from an individual exporter. From the perspective of a European buyer, the attempt to brand South African squid is likely to undermine the reputation of European squid buyers; it may even open the door for a more direct route to European markets that bypasses Spanish and Italian buyers. Not surprisingly, South African exporters report that their packaging is usually removed once it handed over to their European importers.

Upgrading and Development

A key question in recent global value chain research is the relationship between upgrading and development. The relationship is not, however, straightforward: as Bair has argued, it is not clear how upgrading in a firm or sector is related to 'the larger units that are traditionally regarded as the spaces or containers of development, such as the national or regional economy' (Bair 2005, 166). Bair's argument is an important one and highlights the limitations of attempting to draw broader conclusions from single export chains. Yet exploring the relationship between upgrading and working conditions remains an important research question. In the case of the South African squid sector, despite significant improvements in technology and product quality, it is difficult to sustain the argument that upgrading has had beneficial impacts on the crew of squid boats, which represents the largest group of employees in the sector. Indeed, working conditions may have become more difficult with the transformation of the fleet from ski-boats to freezer boats.

The difficult and dangerous conditions facing squid crew was revealed at a recent meeting between the Congress of Trade Unions (Cosatu), the Food and Allied Workers Union, the Department of Labour and around 50 fishermen. At the meeting held in Port Elizabeth workers provided verbal testimony on working conditions, terms of payment and living conditions on the boats. With regard to working conditions, the upgrading of the fleet to freezer boats has allowed boat owners to stay at sea for between 14 and 21 days, depending on the size of the vessel. The squid crew is usually hired on contract, but only for the duration of a single trip. The disadvantage of these short term contracts for the crew is that they are not covered by the legislation that applies to permanently employed workers and which allows them to claim for unemployment insurance, medical aid or pension allowances. Payment schemes are equally problematic for squid crew: wages are determined through a quota system, which is written into the contract. Fishermen must meet the quota – which is set at the break-even level for the boat – before they can expect to be paid. In other words, if the quota is met, all the boat's expenses in terms of fuel, food and equipment is covered. However, if the boat does not catch enough fish to cover the quota – which frequently happens in a sector dependent on a variable resource – 'workers can earn next to nothing, or even end up owing their employers money' (Cosatu 2005). The quota system may explain why an independent survey of squid crew found that many claimed that they often returned from a fishing trip having earned nothing (Charter and Tamae 2004).

The poor living conditions on the boats was an issue raised at a meeting of the organisation representing boat owners, the South African Squid Management Industrial Association. At the 2001 meeting held in Port Elizabeth, representatives from the South African Maritime Safety Authority (SAMSA) reported that their survey of squid boats revealed serious problems in both living conditions and adherence to standard safety regulations. In several cases boats lacked basic safety equipment, which had forced the SASMA official to suspend the boats licence 'for the safety of lives at sea' (SASMIA 2001, 6). Living conditions for squid crew on up to 40 percent of the boats were described as 'unsavoury'. Indeed, the poor state of living facilities on the boats had been reported to the Human Rights Commission offices in Port Elizabeth, which had received 22 written complaints in the 2001 season (SASMIA 2001).

In the last five years there have been important changes in the regulation of squid labour, which are aimed at improving working conditions for squid crew and their living conditions on the squid boats. For instance, one of the new criteria for the allocation of squid quotas is the ability to demonstrate that crew are employed on a full time basis, enjoy safe working conditions and have access to medical aid and a pension. In addition, the country's department of Marine and Coastal Management now requires that all squid quota holders submit a list of crew working on their boats with a view to providing greater job stability for squid workers. Finally, and perhaps most importantly, in 2005 a Statutory Council was established for squid industry workers (RSA 2005). Statutory Councils are institutions that may be established under the Labour Relations Act and operate somewhat like trade unions in that their function is to draw up agreements on wages and working conditions. They can also be used to establish benefit schemes for workers including medical aid and pension. The Statutory Council for the squid industry has established a wide ranging set of guidelines including remuneration, working hours, overtime payment and leave.

These local initiatives to improve the situation for squid crew are important because, based on the evidence presented here, the process of upgrading has not led to improved conditions for the majority of those working in the squid sector. Moreover, the potential for consumer based pressure to improve the working conditions for squid crew may also be limited given that the chain is not (yet) dominated by retailers, who might be placed under pressure to improve the conditions of squid workers by consumer groups in Europe.

Conclusion

Recent efforts to 'upgrade' the value chain framework – by emphasising the significance of social and development issues and by refining the ideas of governance and upgrading – have played an important role in demonstrating the utility of this approach to global trade networks. The South African squid case suggests that the process of 'socially embedding' the framework may be more fruitful than the effort to refine the categories of upgrading and chain governance. Rather than identifying the particular category of upgrading or governance, this chapter explored the trajectory of upgrading over the last 15 years.[1] The early phase of upgrading associated with the introduction of freezer boats had a dramatic impact on the sector: it resulted in the transformation of the squid fleet, the creation of several thousand jobs for

---

1 This descriptive approach to upgrading corresponds to the approach favoured in Gibbon in Ponte (2005, 91) who prefer to analyse upgrading by identifying the 'concrete roles that offer suppliers higher and more stable returns, as well as the routes that they typically use for arriving at them'.

squid fishermen, and it established the local industry as a significant source of high quality squid to the discerning fish markets of Spain and Italy. The more recent phase of upgrading is associated with the auditing of production process, which has become pervasive in agri-food export chains. For squid producers this second phase of upgrading, which they report to be ongoing and relentless, does not generate income gains and is instead a cost that must be borne in order to remain a supplier to European markets. In the face of a very stable export price and a strengthening local currency, exporters are exploring other upgrading options but, for a range of reasons, their efforts have been unsuccessful. The result is that squid boat owners and processor-exporters find themselves continually upgrading but with little impact on their export earnings, while 'real' upgrading appears to be beyond their grasp.

The priority of socially embedding value chains offers an important and policy relevant line of enquiry within value chain research. While upgrading in the squid sector created several thousand new jobs, the new technology associated with freezer boat fishing – together with exploitative contracts – has meant that working conditions in this globally competitive value chain are very poor indeed. Prospects for improving working conditions are unlikely to arise from the upgrading that is currently occurring within the squid industry. On the contrary, the second phase of upgrading is placing additional financial pressures on boat owners; they are unlikely to consider improving working conditions or wages in this context. Prospects for improving working conditions are much better through local regulatory efforts.

References

Bair, J. (2005), 'Global capitalism and commodity chains: looking back, going forward', *Competition and Change*, vol. 9, no. 2, pp. 153–80.

Bair, J. and Gereffi, G. (2003), 'Upgrading, uneven development, and jobs in the North American apparel industry', *Global Networks*, vol. 3, no. 2, pp. 143–69.

Bernstein, H. and Campling, L. (2006), 'Commodity studies and commodity fetishism: trading down', *Journal of Agrarian Change*, vol. 6, no. 2, pp. 239–64.

Charter, S. and Thamae, T. (2004), An Investigation into the Socio-economic Problems Experienced by Chokka Squid Fishers: Port St Francis, Eastern Cape, Norsa 2801 Project, University of Cape Town, Cape Town.

Congress of South African Trade Unions (Cosatu) (2005), Cosatu General Secretary, Zwelinzima Vavi, meet fisherman at FAWU – Port Elizabeth, 25 November 2005, <http://www.cosatu.org.za/speeches/2005/zv20051125.htm> (accessed 10 October 2006).

Gereffi, G. (1999), 'International trade and industrial upgrading in the apparel commodity chain', *Journal of International Economics*, vol. 48, pp. 37–70.

Gereffi, G., Humphrey, J. and Sturgeon, T. (2005), 'The governance of global value chains', *Review of International Political Economy*, vol. 12, no.1, pp. 78–104.

Gibbon, P. and Ponte, S. (2005), *Trading Down: Africa, Value Chains and the Global Economy*, Temple University, Philadelphia.

Gibbon, P. (2001), 'Upgrading primary production: a global commodity chain approach', *World Development*, vol. 29, no. 2, pp. 345–63.

Globefish (2003), *Commodity Update: Cephalopods*, Food and Agriculture Organisation, Rome.

Globefish (2004), *Commodity Update: Cephalopods*, Food and Agriculture Organisation, Rome.

Harilal, K.N., Kanji, N., Jeyaranjan, M.E. and Swaminathan, P. (2006), Power in Global Value Chains: Implications for Employment and Livelihoods in the Cashew Nut Industry in India, Summary Report, IIED, London.

Humphrey, J. and Schmitz, H. (2001), 'Governance in global value chains', in Schmitz, H. (ed.), *Local Enterprises in the Global Economy*, Edward Elgar, Cheltenham, pp. 95–109.

Humphrey, J., McCulloch, N. and Ota, M. (2004), 'The impact of European market changes on employment in the Kenyan horticulture sector', *Journal of International Development*, vol. 16, pp. 63–80.

Nadvi, K. (2004), 'Globalisation and poverty: how can global value chain research inform the policy debate?', *IDS Bulletin*, vol. 35, no. 1, pp. 20–30.

Oceanafrica (2005a), Historical overview and current status of the South African chokka squid fishery, Information System for the South African Squid Fishery, Oceanafrica, <http://www.oceanafrica.com/expertsystem/history.html> (accessed 31 May 2007).

Oceanafrica (2005b), The South African Chokka Squid Fishery, Information System for the South African Squid Fishery, Oceanafrica, <http://www.oceanafrica.com/expertsystem/research.html> (accessed 31 May 2007).

Rammohan, K.T. and Sundaresan, R. (2003), 'Socially embedding the commodity chain: an exercise in relation to coir yarn spinning in Southern India', *World Development*, vol. 5, no. 31, pp. 903–23.

Roel, B. and Butterworth D. (2000), 'Assessment of the South African chokka squid, *Loligo vulgaris reynaudii*: Is disturbance of aggregations by the recent jig fishery having a negative impact on recruitment?' *Fisheries Research*, vol. 48, pp. 213–28.

RSA (2005), Establishment of a Statutory Council, *Government Gazette*, Notice 1573, vol. 482, 19 August 2005, Republic of South Africa.

SASMIA (2001), South African Squid Management Industrial Association, Minutes of the Annual General Meeting, 31 October 2001, Port Elizabeth.

Sauer, W.H.H., Hecht, T. Britz, P.J. and Mather, D. (2003), An Economic and Sectoral Study of the South African Fishing Industry, vol. 1. Economic and Regulatory Principles, Survey Results, Transformation and Socio-economic Impact, Report prepared for Marine and Coastal Management by Rhodes University, South Africa.

Chapter 3

Agribusiness and Poverty Reduction: What Can be Learned from the Value Chain Approach?

Andreas Stamm

Introduction

The Millennium Development Goals (MDGs), adopted by the Millennium Summit of 2000 commit the international community to a strategy of accelerated poverty reduction. The proportion of people living under conditions of extreme poverty should be reduced by half by the year 2015 (MDG 1). The MDGs are today the most significant point of reference for international development policy, cooperation and research (*see* e.g. Fues and Loewe 2005). The commitment to MDG has led to intensified research on the question, which factors influence the poverty reducing effects of economic growth and to the search for growth patterns that have a more direct impact on poverty than mere 'trickle down' processes (Kakwani, Khandker and Son 2004).

One promising sector for pro-poor growth is agribusiness as it allows developing countries to capitalise on their natural production advantages, while reaching out to the rural poor through backward linkages to the farmers. This chapter assesses the relationship between an expansion of agribusiness and poverty reduction, starting with the empirical observation that there is no evident link between a country's specialisation on agriculture and the social welfare of a society. Besides the institutional setting in each country, the social outcome of agribusiness activities is also shaped by the specific way in which local producers are linked to global markets. The value chain approach is an important tool to analyse these relationships. The elements of the approach most relevant in this respect are sketched. This is followed by an overview over the outcomes of empirical research in Sri Lanka's agribusiness. The conclusion introduces some questions that the author proposes as important elements of the future research agenda on value chains.

Agribusiness: A Pillar for Broad-based Growth?

Traditional trade theory suggests an optimistic view of the poverty reducing potential of an agribusiness-based, export-oriented growth pattern. First, under open and undistorted market conditions developing countries can capitalise on locational advantages based on favourable agro-ecological conditions. Second, agriculture tends to be labour-intensive. Increased agri-based exports boosts demand for labour, contributing to rising factor remuneration. Since in most developing countries absolute poverty is concentrated in rural areas, rising incomes in agriculture and related industries tend to have a greater poverty alleviating effect than in sectors that tend to be located in urban areas (manufacturing, services).

A growth pattern based on agribusiness, however, does not *necessarily* lead to pro-poor growth. For example, Honduras has been integrated for decades into the world economy through agricultural products, but, it remains one of the poorest countries of Latin America. In contrast, countries like Australia or Denmark have ascended into the group of high income countries, based largely on internationally competitive agribusiness. Other agribusiness exporters have lost ground in economic development due to a stagnant and mainly rent-oriented agribusiness (Argentina) or have never reached decent living standards (parts of Africa).

This differentiated outcome of roughly comparable development paths can in part be linked to the institutional setting at the national level in which agribusiness growth is embedded. Societies based on a more equitable distribution of the means of production and institutional control of power asymmetries tend to incorporate the impetus coming from a dynamic expansion of agribusiness exports much more broadly than countries characterised by high initial inequalities and an underdeveloped legal control of, for example, transfers in land titles (e.g. Berry 1998).

But it is not only this varying horizontal embeddedness of agribusiness production that makes a difference regarding social and economic outcomes. This chapter argues that it is at the same time the way in which local production is linked to global markets, that explains, to what extent growth leads to poverty reduction. The approach that seems most appropriate to analyse these causal relationships is the value chain approach, especially the line of research that focuses on the governance of chains and thus helps to understand power relations within global value chains.

Agribusiness and Poverty Reduction – The Contribution of the Value Chain Approach

In the past three decades, several approaches have been developed to describe and analyse the structure of production chains and to enable policy makers and other actors to design appropriate intervention in order to enhance economic or social outcome (Stamm 2004; Altenburg 2006). For an analysis of the impact of agribusiness exports on developing country producers, the approach applied and further developed by a group of researchers networked in the Global Value Chains Initiative (http://www.globalvaluechains.org/) seems to be the most promising as it goes beyond intra-firm supply chains (as in the case of Michael Porter's 'Value Chain') and beyond more or less technical analyses of product flows (as in the case of older approaches of production chains or '*filiéres*'). It invites for the analysis of the *varieties* of chain structures based on *common* analytical tools and concepts mainly regarding the power relations within chains and the opportunities of actors to 'upgrade' and thus to stabilise or improve the own position.

The poverty reducing effects of the formation of agribusiness value chains depend mainly on the *governance structures* within the chain and the *barriers to entry* at its upstream ends. Both factors can only analytically be separated from each other, in reality and especially in the dynamic perspectives they are closely interrelated.

In the early work of Gereffi and others (Gereffi and Korzeniewicz 1994) on global commodity chains, there seemed to be a clear-cut division between winners (the industrialised countries of the North) and losers (the developing countries) in the formation of global commodity chains. This lead to particular conclusions. Whether large companies govern value chains through their dominance of production technologies (producer-driven chains, e.g. computers or automobiles) or through their competencies in branding and market access (buyer-driven chains, e.g. clothes or fresh fruit), the relative outcome will be the same: added value will be captured by the countries in the North (location of the lead firms) while the South (location of suppliers and subcontracted companies) will remain as low value-added sites of production and over the years decreasing margins.

Governance Structures and their Effects on Weaker Actors in the Chain

Subsequent research, however makes it clear that the distinction between producer-driven and buyer-driven value chains cannot capture the complexity of new forms of an international division of labour. The distinction among five types of differently governed value chains reflects the efforts to get a more comprehensive picture of the complex situations found in the new international division of labour. So, what can be assumed regarding the socio-economic outcome of the types of value chains mentioned by Gereffi *et al.* (2005)?[1]

In **market-based value chains,** governance, defined as non-market dimensions of the regulation of exchange among economic agents, does by definition not exist. The income that the producer may appropriate, depends basically on market mechanisms, i.e. supply and demand relations. These may significantly be altered by the institutional conditions in the producing countries.

Relational chains are characterised by low levels of power inequalities and by high levels of interaction and information exchange among the chain agents. This implies a rather positive socio-economic outcome. Actors are mutually dependent, no easy 'exit option' for the buyer may allow him to suppress margins at the upper end of the chain by threatening producers to change to other suppliers. Some chains – mainly within the fair trade and organic markets – are regulated by what convention theory denominates 'civic coordination', with actors committed to common values, leading to an intrinsic motivation to avoid conflict (Raikes *et al.* 2000, 18–20).

Captive value chains condense what for a long time was denominated as buyer-driven and producer-driven value chains. The lead firms concentrate on chain functions with high entry barriers and that corresponds to their established core competencies. The producers at the upper end of the chain are often forced into

---

1 We skip discussion of the 'modular value chains'. These can mainly be found in rather sophisticated manufacturing sectors, such as automobiles. A direct link to poverty reduction is difficult to derive.

a fierce price competition among each other and with new entrants to the market. Thus, the income distribution along the chain is highly unequal.

Finally, in **vertically integrated chains** – where all production steps are realised within one single company – distribution of economic outcome is very much the same as in classic capital-labour relationships, complemented by the globalisation dimension. Under conditions of surplus labour, remuneration of (unskilled) workers will be close to the minimum required for survival, determined mainly by the costs of basic food crops. Trade unions as a force to elevate basic wage levels have only developed in few agri-businesses within developing countries; their bargaining power is often being reduced by the permanent threat of companies to relocate production or sourcing to lower-cost regions or countries.

Barriers-to-Entry and their Effects on Weaker Actors in the Chains

Governance of value chains is directly related to the existence or absence of **barriers-to-entry** for producers at its upper end. Captive value chains tend to prevail in sub-sectors of agribusiness with established markets, well known product and process standards and where production technology can more or less be acquired 'off-the-shelf'. Under such conditions entry barriers for producers are low. Lead firms may quite easily identify suitable production areas and induce production, e.g. delivering technology packages and pre-financing initial investment. While this situation may be considered favourable in short term, in the long run it forces producers into a cost-based competition within their countries or with farmers in other parts of the world leading to shrinking margins and increasing levels of vulnerability.

But not in all agriculture-based value chains – and not at every moment in time – barriers-to-entry are low for producers and thus competition merely cost-based. In some chains, areas with the appropriate production conditions are limited. In this case, the producers upstream are in a much better bargaining position. To detect such land scarcities, a sub-sectoral analysis may be required. For illustration: while cinnamon in its generic form may be produced in many countries and thus by a multitude of farmers, its most appreciated variety *cinnamomum zeylanicum* is only produced in some areas of Sri Lanka and South India. Coffee is produced in many countries, however, the conditions to produce high quality *coffea arabica* is limited to highland regions in a limited number of countries.

Higher entry barriers and thus 'scarceness' may also arise from innovation. Often, innovation rents will be appropriated by a small number of actors at the processing or marketing stage of the value chain. However, in many cases, transition towards more specific products (e.g. high value non-traditional fruits) or processes (e.g. corresponding to specific food security standards, Jaffee and Henson (2004)) will modify significantly the governance structure within the chain. In many cases, interaction among the agents increases significantly and the lead firms actively invest in the knowledge transfer towards the actors upstream. Thus, the interrelationships among the players acquire more and more the characteristics of relational value chains.

Empirical Research in Sri Lanka's Agribusiness: Types of Value Chains and Prospects for Poverty Reduction

> It will be important to ensure that future growth of the Sri Lankan economy is more pro-poor than past growth (World Bank 2005, 62).

Sri Lanka is classified as a lower middle income country by the World Bank. The United Nations Development Program (UNDP) Human Development Indicator (HDI) ranked Sri Lanka in 2004 93rd out of 177 countries for which the HDI was calculated; this is below other Asian countries such as Thailand (73) and the Philippines (84), but clearly above, for example, Vietnam (108) and Indonesia (110) (UNDP 2006). The country performs well regarding non-income dimensions of poverty, such as primary enrolment and completion or infant and under-five mortality rates (World Bank 2005, 56).

Poverty in the Sri Lankan context is, thus, in the first instance income poverty. There are huge geographical variations in poverty incidence. It is low in most parts of the Western Province, with numerous employment opportunities in export oriented apparel and other industries and services in the private and public sector. In general, poverty is more severe in rural areas, but even among rural districts variations can be high. A 2005 study came to the conclusion that there is some

> evidence that poverty is concentrated in geographically isolated areas (in terms of distance to markets and cities), the estate sector, and among households with agricultural wage employees (Narayan and Yoshida 2005, 17).

During the 1990s agriculture – and especially Sri Lanka's estate sector – largely stagnated, what can at least partially explain the high level of rural poverty and the fact that consumption inequality has risen considerably during the last decade. Between 1995/96 and 2002 the Gini coefficient rose from 34 percent to 41 percent (World Bank 2005, 12).

Considering these facts but also the conflict potentials arising from inequality and poverty in rural areas (not only in the North-East) the generation of employment and income opportunities in agribusiness has (or should have) a high priority in Sri Lanka's development agenda. The conditions for an increased participation by Sri Lanka in the global agribusiness market are rather good, as the country has favourable agro-ecological conditions for a diversified agriculture and the production of high value crops. Since colonial times, the country has been integrated into the world market with agriculture based goods, mainly tea, rubber, coconut and its derivates, and spices.

Empirical research was conducted in Sri Lanka in 2004. Around 100 interviews with key stakeholders of a variety of agri-based value chains – from plantation crops to minor crops, such as spices, essential oils and specialty organic agribusiness products – were conducted. Interviewees included various economic actors along the chains, from farmers to processors, exporters and retailers. Following a systemic approach to competitiveness analysis, interviews were also conducted with representatives of institutions on the meso-level (research institutes, business associations and so on) and with policy makers at the macro level. The following section sketches the outcomes of this research in a summarised way. (More details can be found in Stamm *et al.* 2006).

Economic Agents within Sri Lanka's Agribusiness

Sri Lanka's agribusiness is mainly made up of three types of players: peasants, collectors/middlemen and exporters, who also do the processing of crops in the cases that this is required.

The land tenure structure in Sri Lanka is clearly dominated by small farmers. The average size of land owned is 0.6 ha; about 71 percent of agricultural households cultivate less than 1 ha, 90 percent less than 2 ha (World Bank 2003, 23). In 2002, only 19.9 percent of all agricultural land was composed of units above 20 acres (8.09 ha). On the main part of this 'estate land' (65 percent) plantation crops such as tea, rubber and coconut are produced; state ownership is still common.

Collectors and middlemen are the link between farmers and processors/exporters. Middlemen may be farmers themselves or independent actors, who receive all their income from collecting and trading goods. Sometimes middlemen fulfil a more comprehensive role for farmers, e.g. providing short term credits.

The main part of processing of agricultural products is done by exporting companies. The agribusiness export sector is rather broad, consisting of around 300 companies. Our interviews in this sector showed a low degree specialisation at the company level. Most firms offer a wide range of agri-based products and are sometimes also active in areas such as importing or tourism. Many companies have roots dating back decades and a long history of trading with partners from the Middle East and Great Britain. Muslim ownership is common among trading companies.

Types of Value Chains in Sri Lanka's Agribusiness

Following the classification of value chains by Gereffi *et al.* (2005) as the conceptual starting point, our research aims to identify types of chains present in Sri Lanka's agribusiness and evaluate their effects on the socio-economic conditions of the weakest players, i.e. small farmers and workers.

We found very few value chains that might be categorised as captive, i.e. chains with a dominating lead firm that sets and controls all important parameters. International retailers or food processors do not appear as important players, in the first instance, due to a lack of scale. Retail companies from OECD countries will only invest in the establishment of supply chains, if the location guarantees sufficient produce to be sourced. The small territory of Sri Lanka limits these quantities and so does the dominant smallholder structure, characterised by multicropping farms with low inputs of fertilisers and other inputs that might increase yields. The same lack of scale also hinders the establishment of processing plants that require high initial investment. We found for instance, that Sri Lankan pepper – often with very high contents of essential oil – is shipped to India to be processed there, together with (often lower quality) pepper harvested in larger scale in the neighbouring country.

Based on our interviews, four basic types of value chains were identified. Listed in ascending order of integration they were coined as 'disintegrated', 'weakly-linked', 'relational' and 'vertically integrated' value chains. To simplify the analysis, we will treat the first two chain types together.

Disintegrated and Weakly Linked Value Chains

The majority of Sri Lankan exporters have very little direct contact with the final buyers of their products, the retailers or processors instead they trade with brokers in the final markets. This situation of chain disintegration is common for most exporters supplying to non-OECD markets. In the case of tea and some spices, transactions are coordinated through auctions, leading to some kind of weak chain integration.

For exporting companies, low chain integration has some advantages: They can react flexibly with regards to product range and quantities. Costs related to the management of one's own plantations or of outgrower schemes can be avoided. While disintegrated chains are an appropriate way of supplying bulk products to markets with rather low entry barriers, they show serious deficiencies when markets are more demanding:

- A stable product quality cannot be guaranteed, due to a poor feedback between exporters and producers;
- Traceability of products up to the farm level, increasingly required on OECD markets, is impossible in disintegrated value chains;
- Low value chain interaction means weak supply coordination, making it difficult to meet requirements related to right-on-time delivery or constant quantities and quality;
- Knowledge about quality standards, prices or market trends do not reach the upper ends of these chains.

Auctions as instruments of weak value chain integration may help to overcome some of these problems. Traceability can be assured, as the origin of the produce is labelled on each batch to be traded. Basic feedback between buyers and producers regarding quality aspects is given. However, auctions are generally characterised by little direct contact between the partners, minimising knowledge transfer among the agents and possibilities to start joint efforts of product and process upgrading.

Strongly Linked, Relational Chains

While the majority of the value chains in Sri Lanka are rather fragmented, there are some in which the agents are linked in a systematic way through longer-term commercial relations, intense interaction, mutual information and knowledge transfer. These relational chains can be found where the business objective is to supply high-value products to challenging markets.

Relational chains are common in the fair trade and organic markets. Here traceability up to the farm level and value chain transparency are not only an advantage but a condition for market access. The system relies on the consumers' willingness to pay a mark-up on market prices, being assured that the goods they buy are produced without inputs that may harm their health and the environment or produced under conditions that guarantee producers and workers a decent living wage.

In Sri Lanka, the most important mechanism to coordinate relational chains are outgrower schemes. They link small farm units to buyers that provide production and marketing services, inputs, knowledge and technology. The smallholder is responsible for growing and sometimes for the first steps in processing, whilst the contractor does further processing, marketing, and manages the whole operation. For the contracting companies, outgrower schemes have a series of advantages compared to other forms of value chain organisation:

- They allow thorough supply-base management and development. The high level of interaction assures good and stable quality of produce;
- Produce is easily traceable up to the farm level;
- Training and knowledge-transfer lower considerably the level of post-harvest losses.

However, relational chains characterised by outgrower schemes also carry some disadvantages and risks for the contracting company. Building up such interlinked chains involves a significant long-term investment for a contractor that will only pay off after several years. The basic reason for this is that in the first years of operation, transaction costs are high, while the quantity of produce sourced through contract farming is still low. Transaction costs of building up a contract farming system may easily transform into sunk costs, if the establishment of outgrower schemes fails.

Considering these costs and risks, two empirical observations from the field studies in Sri Lanka are noteworthy:

- The integrators of relational chains do not belong to the smallest group of enterprises, but rather to the medium to large companies, that can afford the initial investment costs and can take the associated risks;
- Many of the companies operate in market segments that lower the uncertainty of international transactions and permit the contractors, to gradually develop their operations, such as the organic and the fair trade market.

Vertically-Integrated Chains

These chains are characterised by actors on the different steps of value chains that operate under the same legal personality or in companies with the same dominant capital owners and thus under a common mandate. Vertically-integrated value chains represent the highest degree of interaction among the agents. For the management of the export business, vertically integrated value chains offer some of the advantages mentioned for the case of the relational chains, namely the traceability of the products and the possibility for upgrading efforts involving different steps of the value chain.

An additional advantage of vertically integrated chains is that they usually include large scale operations, such as plantation farming. An effective division of labour can be introduced, and production and transaction costs greatly reduced. Implementing new rules and changing production are much easier to achieve than in fragmented chains. This allows for smooth operations of vertically integrated chains and also adds to the direct control of the supply base. However, in the case of Sri Lanka, few vertically integrated value chains were identified, mainly because a rather restrictive land policy impedes the concentration of larger extensions of land.

Social Inclusiveness and Potentials for Poverty Reduction

The dominating disintegrated value chains have severe shortcomings not only with regard to the requirements of international markets but also with regard to social inclusiveness. These shortcomings are related to power and information asymmetries and to the absence of learning and upgrading processes.

Due to the highly scattered production structure and the lack of effective producers' organisations, smallholders are dependent on middlemen to take their produce to the market. While middlemen fulfil necessary functions within agribusiness and often enjoy the confidence of the farmers, in other cases they reap farmers off a considerable part of the value, mainly where middlemen are the only option for farmers to get their produce to the market. Power asymmetries are especially high in the case of perishable products, because farmers can not wait for other intermediaries to appear. Unequal distribution of bargaining power is further aggravated by information asymmetries, due to a lack of access to market information by the farmers.

The position that farmers have in disintegrated value chains cut them off from important information emanating at the lower end of the chains, i.e. at the point where exporters receive valuable information about market trends with regard to consumer preferences, standards to be introduced or price trends. This prevents farmers from systematic learning and upgrading processes that might improve their position within the value chain and assure or increase future income.

The overall situation of the smallholders is more favourable with regard to relational chains. At first sight the position of the exporter/processor that integrates relational value chains through outgrower schemes is very much alike that of the middlemen regarding aspects of information and power asymmetries. However, there are significant differences that raise the bargaining position of the farmers and thus increase their benefits from the transaction in the short and in the long run:

- The establishment of relational value chains implies specific investment by the lead firm, basically in training of and knowledge transfer to a defined group of farmers. Once this investment is realised, its value is incorporated in these actors, augmenting their bargaining power. A break-up of the relationships between farmers and lead firms would inevitably lead to the latter having to sacrifice sunk costs and even worse, competing companies might be free-riding on the initial investment;
- Relational chains are characterised by intense bi-directional information flows. This does not imply that problems of information asymmetries are absent in this kind of chains, because the lead firms will tend to be selective in the kind of information that they pass on to the farmers. However, in many cases the knowledge transferred in order to guarantee smooth functioning of the chains

can induce learning processes beyond any strategic control by the lead firm;
- An additional factor that clearly favours the position of smallholders in relational chains is that the lead firms tend to foment the organisation of the farmers in order to lower transaction costs and to realise economies of scale in the area of post-harvest handling, processing or packaging of produce. Organised producers are most often in a much better bargaining position than individuals.

Being part of a contract farming system does not always imply a very significant net increase in the income of farmers' households, at least in the cases of outgrower schemes visited by our research team. Most of the farmers work on small and diversified holdings. Major parts of the farms are subsistence oriented, while the outgrower schemes only apply to one or few commercial crops. While the average increase in income is limited, a major poverty alleviating effect results from an increased security regarding the cash income received during the harvest periods. The prices are fixed in advance and purchase by the contractor guaranteed. This lowers the vulnerability of poor farmers' households considerably.

An important dynamic aspect of social inclusiveness in the case of the analyzed relational chains are the learning processes induced by training and technology transfer organised or paid for by the lead firms. Much of the acquired know-how that in first instance applies only to the crops that are of interest for the contracting firm can be transferred to other parts of the smallholders' production (e.g. composting techniques), increasing productivity, reducing fluctuation and enhancing ecological sustainability of farming.

Vertically integrated value chains are not often found in Sri Lanka, and when are, largely in the estate sector (tea, coconut, rubber). All three estate products suffered for many years from structural oversupply and thus adverse price trends. This lead to a very vulnerable position for the workers and dependent tenant farmers. Thus, poverty is a very serious problem in the estate sector, comprising different dimensions of poverty: low income, poor access to safe sanitation and drinking water, cooking fuel and electricity.

The situation is different in tea value chains that are certified as providers to the fair trade market segment. In order to be eligible for the Fairtrade register, tea producers as well as other companies depending on hired labour must guarantee the compliance with a series of social and labour standards (e.g. the ILO core labour standards) and have to invest the fair trade premium in development projects designed jointly by plantation management and representatives of the workers.

Final Considerations and Open Research Questions

Value chain research helps to understand, why growth based on export oriented agribusiness may have very different social implications. Thus, it can be an important tool for policy makers in developing countries and for development cooperation. Even if value chains cannot be 'constructed', governments and donors can set incentives and improve framework conditions by preference for the formation of competitive value chains with a high impact on poverty reduction. In the best case relational chains can have an important outreach to smallholders and especially where the chain integrators are willing to share their knowledge and enable joint learning and upgrading.

The concept of governance of value chains, a very important contribution to the understanding of international production and exchange of products, is still 'work in progress'. Gereffi *et al.* (2005, 96) indicate, that governance patterns of value chains are 'neither static nor monolithic'. Our research in Sri Lanka supports the idea, that governance structure shift with time. Relational chains prevail where know-how has to be created among various players or where intense interaction is required to develop marketable products in demanding markets, e.g. in the organic agribusiness. However, once the know-how is consolidated and easily transferable to other players (in our case farmers) relationship among agents may acquire more characteristics of captive value chains.

A similar situation can be found with regard to international standards. Under a situation where few companies (exporters or processors) adhere to, for example, Hazard Analysis and Critical Control Points processes, buyers on the international markets will be willing to maintain carefully managed relations with them, in order to guarantee access to products that fulfil the (new) requirements. Once external governance makes compliance with certain standards more and more ubiquitous, the same processors or exporters may become easier to be substituted – and the chain may become more of the captive type. Thus, future research may focus on the *dynamic* dimension of value chain governance and the interrelationship between external and internal governance.

Acknowledgements

Andreas Stamm would like to thank two peer reviewers for important comments to a first draft of the paper.

References

Altenburg, T. (2006), Donor Approaches to Supporting Pro-poor Value Chains, The Donor Committee for Enterprise Development, Technical Paper <http://www.enterprise-development.org>.

Berry, A.R. (1998), When do Agricultural Exports Help the Rural Poor? A Political-Economy Approach, University of Toronto, Department of Economics, Working Paper Number UT-ECIPA-BERRY2-98.

Fues, Th. and Loewe, M. (2005), 'Between Frustration and Optimism: The Development Outcome of the Millennium+5 Summit', German Development Institute, Briefing Paper No. 7/2005, Bonn<http://www.die-gdi.de/die_homepage.nsf/56a1abebb4eded3dc1256bd9003101ea/90f39efa0fed80bec1257092006337f1?OpenDocument>.

Gereffi, G., Humphrey, J. and Sturgeon, T. (2005), 'The governance of global value chains' *Review of International Political Economy*, vol. 12, no. 1, pp. 78–104.

Gereffi, G. and Korzeniewicz, M. (eds) (1994), *Commodity Chains and Global Capitalism*, Praeger, London.

GoSL (Government of Sri Lanka) 2002, *Regaining Sri Lanka: Vision and Strategy for Accelerated Development*, Sri Lankas Poverty Reduction Strategy Paper, Colombo.

Jaffee, S. and Henson, S.J. (2004), *Standards and Agri-food Exports from Developing Countries: Rebalancing the Debate*, World Bank Policy Research Working Paper 3348, World Bank, Washington D.C.

Kakwani, N., Khandker, S. and Son, H.H. (2004), *Pro-Poor Growth: Concepts and Measurements with Country Case Studies*, United Nations Development Programme, International Poverty Centre, Working Paper No. 1, August 2004, Brasilia.

Narayan, A. and Yoshida, N. (2005), *Poverty in Sri Lanka: The Impacts of Growth with Rising Inequality*, World Bank, South Asia Region, PREM Working Paper Series, Report No. SASP PR-8, World Bank, Washington D.C.

Raikes, P., Jensen, M.F. and Ponte, S. (2000), *Global Commodity Chain Analysis and the French Filière Approach: Comparison and Critique*, CDR Working Paper 00.3, CDR: Copenhagen.

Stamm, A. (2004), Value Chains for Development Policy: Challenges for Trade Policy and the Promotion of Economic Development, Eschborn. <http://www.die-gdi.de>.

Stamm, A. *et al.* (2006), 'Strengthening Value Chains in Sri Lanka's Agribusiness. A Way to Reconcile Competitiveness with Socially Inclusive Growth?' DIE-Studies 15, Bonn.

UNDP (United Nations Development Programme) (2006), *Human Development Report 2006*, United Nations, New York.

World Bank (2003), Promoting Agricultural and Rural Non-Farm Sector Growth, World Bank, Colombo.

World Bank (2005), Attaining the Millennium Development Goals in Sri Lanka: How Likely and What Will It Take To Reduce Poverty, Child Mortality and Malnutrition, and to Increase School Enrolment and Completion? The World Bank Human Development Unit, South Asia Region, Washington D.C.

Chapter 4

Big is Not Always Better: Global Value Chain Restructuring and the Crisis in South Indian Tea Estates

Jeffrey Neilson and Bill Pritchard

Introduction

The restructuring of the tea, coffee and cocoa sectors has provided a rich source of material for research into the connections between global value chains and international development and poverty reduction. An extensive body of literature has documented the inter-relations between these chains being transformed from state-centric to global-private in their organisational logic, and the coincident crash of global prices (Baffes 2005; Daviron and Ponte 2005; Fold 2001; Gibbon and Ponte 2005; Gilbert 1996; Neilson 2007; Ponte 2002; Talbot 2002). These research contributions, firstly, have documented how the nascent regime of global private-regulation has been authored largely by an influential cadre of lead firms – supermarket chains and multinational food companies (coffee roasters, tea blenders and confectioners) – whose commonality relates to the imperative to possess, extract value and protect valuable intangible assets stored in retail brands; and secondly, have identified the processes of value chain re-regulation as lead firms seek to attach various product and process attributes to the raw and semi-processed products they purchase, thereby incorporating upstream producers within modes of ordering based around traceability requirements and harmonised grades and standards (Busch and Bain 2004; Ponte and Gibbon 2005; Reardon *et al.* 2001).

However, significant questions remain to be answered in terms of 'what this means' for production structures in developing countries. Mostly when this question is asked, the focus is on the 'smallholder problem'; namely, the issues of smallholder exclusion and/or marginalisation from global value chains. Yet while such a focus appropriately responds to agendas relating to poverty alleviation and rural development – given the highly vulnerable role of smallholders in the rural economies of developing countries – it tends to divert attention from what is happening within the estate sector, which traditionally has accounted for the bulk of output and employment in many key production sites. This chapter seeks to address this research lacuna.

To the extent that the estate sector has figured in recent research, the abiding view has tended to be that, if anything, global value chain restructuring will generate some competitive advantages for estates, vis-à-vis smallholders, because these entities are generally better equipped to respond to lead firms' demands for certification, traceability, and the delivery of relatively larger volumes of standardised output that meets increasingly specific buyer requirements. With very limited recent research evidence on how the estate sector in the tropical products industry has responded to contemporary value chain restructuring, these arguments are informed largely via the well-documented processes of change in the emergent export fresh fruit and vegetables (FFV) complexes across a swathe of developing countries in Africa, Latin America and Asia servicing export markets, and domestic and foreign supermarkets (Barrett *et al.* 2004; Birthal *et al.* 2005; Dolan and Humphrey 2000; Friedberg 2003; Gibbon and Ponte 2005; Hu *et al.* 2004; Reardon and Timmer 2007; Weatherspoon and Reardon 2003). By and large, evidence from FFV suggests that larger producers are the beneficiaries of recent changes, as smaller 'less-connected' producers become marginalised and there is a rapid consolidation of production into larger supply enterprises. On this basis, it might be expected that the estate sector does not constitute a 'development problem' and, indeed, might offer a 'development solution' to the question of how developing countries respond to global value chains (Humphrey 2006).

Using evidence from the tea estate sector of South India, this chapter complicates such an interpretation. It argues that contemporary processes of global value chain restructuring will not unambiguously favour larger production units. Restructuring within the South Indian tea sector is occurring within a globally depressed world market and amidst continually increasing domestic costs of production. The world tea market is characterised by stagnating consumption and ever-increasing production, particularly in China, Vietnam and Kenya; a situation expected to continue in the medium term (FAO 2006). In 2005, for the first time in the modern era, tea production in China surpassed production in India to become the world's leading producer. At the same time, consumer tea markets are applying increasingly strict quality requirements related to food safety, ethical standards and traceability (Neilson and Pritchard 2006). These conditions, and intensified global competition, are generating serious profit crises for the tea estates of South India. Estate owners are currently seeking to repair their problematic positions through a variety of restructuring initiatives. As a consequence, the traditional 'estate model' is being compromised and challenged. Whether these estates can be reconstituted into new, economically viable and socially desirable forms, is not at all certain at this stage. Future trajectories of industry development will depend upon the outcomes of various political and economic struggles currently taking place both within the plantation districts of South India and across the global value chains of which these regional economies are embedded.

The analysis of the industry presented here is based on field research carried out between 2004 and 2006 with the cooperation of the United Planters' Association of Southern India (UPASI). Interviews were held with approximately 100 informants, including plantation and factory managers, agricultural extension officers and other bureaucrats, smallholders, representatives of industry associations, union leaders, and workers. Informants were commonly identified through consultation with key actors such as UPASI, the Tea Board of India and NGOs. Interviews were semi-structured around key themes identified in accordance to the role and position of the informant. These interviews were then complemented by the examination of secondary sources including industry bulletins, consulting reports and other 'grey' literatures. An extended discussion of this research will appear in Neilson and Pritchard (forthcoming).

Tea in South India

Tea was first grown in South India by the British in the late 19th century, with production essentially confined to large-scale, private estates until Indian independence in 1947. Today, the South Indian tea industry is spread across a string of planting districts in the States of Kerala and Tamil Nadu in the Western Ghats (Figure 4.1). In 2002, estates accounted for 55 percent of South India's tea production, with the remaining smallholder production concentrated almost exclusively in the Nilgiri Hills of Tamil Nadu (Tea Board of India 2003). Tea is commonly the mainstay of the regional economy for the districts within which it is grown, and employs an estimated 360,000 people across South India (Tea Board of India 2003). Tea estates are frequently enclave economies in otherwise remote and isolated areas of the Western Ghats, where management provides basic services and infrastructure for the community, including schools, health care, and roads. In a general sense, the industry forms part of a broader plantation crops economy, also including coffee (grown mainly in districts further north, in Karnataka), rubber, and spices including pepper, cardamom and vanilla.

Total South Indian tea production averages 200 million kilograms annually; substantially less than the 650 million kilograms produced in the North Indian states of Assam and West Bengal (Tea Board of India 2005). However, tea exports from South India (averaging around 100 million kilograms in recent years) actually exceed North India exports, such that the export-oriented South Indian sector is far more susceptible to shifts in global tea markets. South Indian tea generally fetches a lower price than North Indian tea, and this price differential has tended to increase over time (Sanjith 2004). The major export destinations for South India tea continue to be the Middle East and Russia (together accounting for 70 percent of total exports in 2002), although exports to Russia in particular have declined significantly since the 1990s.

The vast majority (85 percent) of South India tea is manufactured using the CTC (Cut, Tear, Curl) method (Tea Board of India 2003), which yields more cups of tea per unit of weight than the other common method of preparing black tea, know as the Orthodox method. Orthodox teas, however, have generally been more saleable on world markets in recent years. It is essential that fresh tea leaves are processed soon after plucking, and factories are necessarily located in close proximity to the plantations. Almost all large estates in South India manage their own factory, although Bought Leaf Factories (BLFs), independent from a plantation base, are also widespread in centres of smallholder production.

Despite their attractive visual appearance, most tea plantations are extremely industrialised production units. Relatively high levels of fertilisers are required to

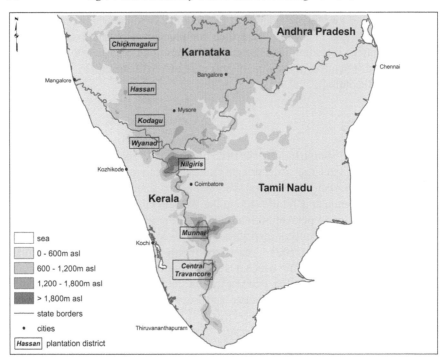

Figure 4.1   Plantation district

Source: Authors with assistance from Jasmine Glover.

replace nutrients lost to leaf removal, and pesticides are applied to mitigate the inherent ecological dangers of a cultivated monoculture. Pruning cycles and plucking regimes are closely monitored and, despite some attempts towards mechanisation in South India, labour costs continue to account for some two-thirds of total estate expenses.

The present-day tea industry in South India has inherited a dense organisational framework from its colonial forebears. Each of the major coffee (and later tea) districts of South India, established industry representative bodies, Planters' Associations, which, by 1894, established joint representation through the United Planters' Association of Southern India (UPASI). These organisations continue to exert a considerable political lobbying force. They also represent management in minimum wage negotiations, undertake their own research and development activities, and increasingly engage directly with global markets. The persistence of exclusive planters' clubs, sporting events such as cricket and golf tournaments, and annual social dinners, are further testament to the strong legacy of British culture on the industry. In addition to the numerous, mainly family owned and operated estates, the tea estate sector in South India today is also comprised of a variety of corporate forms, from dedicated plantation companies to diversified conglomerates and trading companies.

Overlaid on these colonial foundations is the equally opaque edifice of post-colonial Indian bureaucracy. The Tea Board of India came into being in 1953, as a statutory body under the Ministry of Commerce, with its headquarters in Kolkata and a South Indian Zonal Office in the Nilgiris District of Tamil Nadu. Whilst the privately owned estates were never seriously threatened with nationalisation, and plantation crops were exempted from land reform under the various State-administered Land Ceiling Acts, the Nehruvian belief in strong state control left its mark on the tea sector. The Tea Board of India has considerable legal power to intervene in virtually every aspect of the industry's operation, including price control, quality, distribution and marketing. At various times, it has used this authority to impose requirements, such as the compulsory use of domestically-manufactured tea chests, forced sale of tea through domestic auctions, and restrictions on foreign ownership and the employment of expatriates on estates.

Another feature of the post-independence Indian tea industry has been the rise in influence of labour unions. The Estates Staff Union of South India (ESUSI) was established in 1947, and has since been eclipsed in importance by the political party aligned unions, such as the Centre of Indian Trade Unions (CITU), the All-India Trade Union Congress (AITUC), and the Indian National Trade Union Congress (INTUC). Labour-management relations on the estates are regulated within the broad framework of the 1951 *Plantation Labour Act*, which obligates estate-owners to provide scheduled levels of social welfare provision for their workforces, including the funding of housing, schools, hospitals, childcare facilities, and the provision of food staples. Intermittent periods of labour unrest and conflict have occurred on the South India tea estates, notably in Kerala, where a Communist government was first elected in 1957 and has held power for much of the intervening years (most recently being re-elected in 2006).

Trajectories of Restructuring in South India

The narrower terms of this chapter are to provide an overview of the key elements of this industry in flux. There is considerable restructuring, uncertainty and experimentation occurring within the estate sub-sector presently. Although the ultimate direction of these developments has not yet fully crystallised, at least four general strategic trajectories can be distilled from current trends: (i) attempts at intensification and productivity gains; (ii) flights to quality; (iii) the recomposition of ownership arrangements, and (iv) estate abandonment. Taken together, these trends emphasise the fundamental point of this chapter, which is that the fate, and assumed superiority, of larger productive enterprises is far from assured in contemporary global value chains.

Intensification and Productivity Gains

Strategies of intensifying tea production (thereby seeking to resolve profit crises through reducing costs per unit of output) potentially incorporate a wide array of agendas. The intensification imperative is most starkly apparent in the case of labour relations, where industry has attempted to freeze any increases to the base wage rate for pluckers (payable once workers pluck a daily minimum of 16kg of green leaf).

In Kerala, five-yearly wage rounds saw base rates successively increase through the 1980s and 1990s. Then, the post-1998 crash of tea prices stimulated an environment where management could implement wide-ranging reforms to wage negotiations. No increase to the base rate had occurred in Kerala since 1996, when, in 2005, Unions and the Kerala Government (in the context of a looming State election) re-opened these negotiations and proposed a 15 percent increase in the daily base rate. It is clearly unacceptable, and ultimately unsustainable, to achieve enhanced efficiency through depressing wages in what is already a very low-paid vocation. Management, however, has argued that the industry could not afford a wage hike under current conditions, pointing out that although base rates have not been increased, the level of take-home pay for pluckers has actually risen significantly. A number of estate companies have developed their own wage incentive regimes over and above scheduled minimum payments. Restructuring of the tripartite wage-setting arrangements between unions, state governments and industry bodies appears inevitable, as estates shift toward company specific wage setting mechanisms, linking wages to productivity gains.

Because of the crisis, industry lobby groups successfully obtained an exemption of Agricultural Income Tax payments in Kerala, and the abolition of this tax in Tamil Nadu. Over the period 2002–2007, the industry has also been provided with financial incentives for productivity enhancement from the Tea Board of India. These incentives include subsidises for replanting and pruning, the creation of irrigation facilities, for conversions of factories from CTC to orthodox manufacturing systems, and for obtaining HACCP certification.

Individual estates in South India have also experimented with significant increases in mechanisation. The use of hand shears (or at least shifting from hand-plucking to shear-plucking throughout the pruning cycle) is now widespread in Kerala, where rural wages remain higher than in other states, such that some estates have halved their workforce. Other estates have attempted to introduce machine harvesters which, however, require uprooting, substantial earthworks, terracing of tea gardens with suitable row spacing, and replanting. The introduction of subsoil drip irrigation has also increased yields considerably, particularly in the drier months, helping to spread production throughout the year. Computerisation has been used to store field data and therefore optimise plucking cycles, and whilst most tea factories rely on out-dated machinery, some are increasingly automated and streamlined. One highly computerised factory in Gudalur District claims to have increased unit production per factory worker threefold due to the introduction of various labour-saving devices and extensive computerisation. Further labour reductions on the South Indian tea estates appear inevitable, as estates find it increasingly difficult to attract rural workers, with many unskilled labourers preferring to work elsewhere, such as in India's booming construction sector.

Taken together, these actions comprise an elaboration of the productivity treadmill in the sector. They represent the imposition and application of the most recent technologies and management practices to the task of growing, plucking and processing tea as efficiently as possible.

Flights to Quality

One widely recommended pathway through which producers can avoid a 'race to the bottom', and ever-declining commodity prices, is the production of high-quality teas for specific markets: the so-called 'quality turn' (as discussed in Goodman 2003). However, quality markets are fraught with danger and do not always guarantee improved returns in the long term (Daviron and Ponte 2006; Neilson 2007). Furthermore, quality in the food and beverage market frequently must be constructed and promoted to become an economic advantage (Renard 2005), a process which is all too often controlled by branded food companies at sites of consumption rather than by producers (Neilson 2005; 2007).

In the South Indian tea sector, declining quality is frequently identified as a primary cause of low prices and falling profits for producers. Whilst this low quality scenario may be applicable at a broad industry-level, several individual estates have managed to establish strong reputations for high-quality teas in niche markets. Implicated within these quality agendas has been an overall tendency to by-pass auction channels in favour of direct sales. Predominantly, teas sold directly are those associated with quality attributes, and fetch relatively higher prices, which has left the auction system increasingly, though not always, as a repository for teas of last resort.

Large estates are admittedly better positioned than smallholders to both produce high quality tea and to present these quality attributes in international markets. A number of high profile estates in the Nilgiris district are able to routinely demand prices three to four times prevailing prices in the auctions at Coonoor, Coimbatore and Kochi. In contrast to the other tropical beverage crops (coffee and cocoa), tea commonly exits the estate as a ready-to-consume product, and various estates sell branded, packaged tea in both the domestic and international markets.

In South India, regional branding of the various planting districts has become an important tool for producers to capture the value of quality associations. This is perhaps most pronounced in the Nilgiris, where the Nilgiri Planters' Association (NPA) has initiated a move to certify member estates as conforming to minimum quality and ethical requirements. It is envisaged that the economics of this initiative will operate like a Geographical Indication, but exclusively for the District's estate sector. UPASI, with the support of the Tea Board, has also organised the 'Golden Leaf India Tea Awards', incorporating an international jury to reward high quality teas from South India. So far, competitions have been held in Coonoor and Dubai. The NPA also managed a tea auction at the 2006 World Tea Expo in Las Vegas, to showcase the region's quality teas. These initiatives reiterate the proactive stance taken by the industry to present quality attributes in international markets. It needs to be pointed out, however, that the market opportunities offered by niche quality tea markets remain ultimately limited in scope and are unlikely to provide a solution for the majority of tea estates.

Aligned to this 'flight to quality' is a wider set of global processes relating to value chain governance and traceability. Some estates (and BLFs) have gained access to premium prices through certification schemes, such as fair trade and organic. From the perspective of branded tea companies, it has certainly become increasingly important to ensure authoritative traceability trails to upstream suppliers. As discussed in a wide body of literature elsewhere (see the discussion in Hughes 2005) these imperatives derive from various combinations of food safety legislation, corporate social responsibility agendas, and corporate imperatives to protect brand equity. Their footprint is to inscribe audit cultures within chains. This, in turn, allows these distant actors to control the means of quality construction within sites of production, and problematises producer attempts to benefit from quality differentiation. As we contend elsewhere: 'standards and certification schemes are more than mere neutral market lubricants: they are strategic tools for supply chain governance which can alternately be empowering or constrictive for rural producers' (Neilson and Pritchard 2007).

Recomposition of Ownership Arrangements

A most striking example of ownership restructuring has taken place in the Kanan Devan Hills[1] of Kerala, around the Munnar hill station. Following Indian independence, estate ownership in Munnar rapidly consolidated with one company, James Findlay and Co., eventually coming to possess 17 out of 23 estates. In 1976, when Findlay quit India, these estates were acquired by the Tata Group, who thus became regional economic patriarchs. The Tata Group is a sprawling Indian conglomerate, headed by the Tata family, who belong to India's small Parsee community. Majority ownership of the group is held by the various charitable Tata Trusts, and the Tata family professes to bring its own brand of social responsibility and philanthropy to its corporate empire.

The estates acquired by Tata covered approximately 24,000 hectares (8,900 ha actually planted with tea) delivering annual outputs of approximately 20 million kilograms of made tea. These activities supported a workforce of approximately 15,000 and a dependent district population of around 50,000. By 2000, the Munnar estates occupied a key flagship position within Tata's South India plantation interests,[2] but were delivering sustained financial losses, equivalent to 2 percent of annual turnover.

Tata Tea was a vertically integrated company which owned estates, processing plants, blending operations, and retail brands. Significantly, whereas the production (estate) side of the operations was losing money, the tea division in its entirety remained profitable. A substantial proportion of the tea coming out of Munnar was used for Tata's 'Kanan Devan' brand, a nationally-recognised, and strongly place-informed, product. Kanan Devan tea was grown, blended and packaged in Munnar, and was a potent geographical symbol for the throng of domestic tourists who regularly visit the town. Effectively, this presented a business case in which, should Tata walk away from Munnar, its valuable domestic tea brands could have their reputations damaged.

---

1 The name 'Kanan Devan' reputedly came from the names of two Muthuvan tribals, 'Kanan' and 'Devan', who assisted the first British Colonials to traverse the forest undergrowth (the effect of which, ironically, was to lead to their own dispossession).

2 In addition to these tea operations, the Tata Group through a different subsidiary also owned and operated a number of coffee plantations in Karnataka.

The business case logic of seeking to maintain a tea supply out of Munnar (though not necessarily owned by Tata itself) dovetailed with the Group's philanthropic reputation. In 2004, various negotiations resulted in what has been labelled 'the world's largest employee buyout'. Some 96 percent of estate workers purchased shares in a newly constituted company, Kanan Devan Hills Plantation Co. (KDHP). The median share purchase was about US$78 per worker, with a lock-in period of three years (afterwards, they are tradable only to other shareholders). Through these arrangements, US$3 million of new equity was raised, which was used as the basis for further loans from a commercial Bank (worth US$5.55 million) and a subordinated loan from Tata (worth US$7.78 million). For these monies, KDHP obtained a sub-lease on all Tata's estates in the Munnar leasehold area, plus all fixtures (factories, offices, bungalows, etc).

Interpreting this chain of events raises a series of issues, not all of which remain resolved. The first and most obvious point is that, to the time of writing at least, these processes seem to have delivered a potentially viable future for the workforce of this plantation district. Tata has exited the estate sector, but has done so in accordance with a new institutional structure that provides for the continuation of managerial and productive functions on the estates.

Without detracting from the legitimate elements of social philanthropy, the 'Munnar model' equally serves Tata's own financial interests. It has provided a route out of a loss-making enterprise and contributes to the Group's overall emerging focus in brand-centred business activities, recognising the increasing profitability of intangible assets in the agri-food sector (refer to Pritchard 2000). Related to this point is the vexed question of the 'Kanan Devan' name. Tata has retained ownership over this trademark, and still runs its blending and packaging operations just outside of the town. Moreover, the subordinated loan provided by Tata to KDHP (with an interest rate of 8.5 percent repayable quarterly) provides an additional income stream for the group.

Therefore, whilst the establishment of KDHP could be interpreted as providing a lifeline for this district economy, alternatively it represents the localisation of responsibility. In its simplest sense, the Tata-KDHP transaction can be understood to have involved an impoverished workforce taking ownership of a loss-making enterprise in an attempt to save their own futures. Interestingly, however, the finances of the enterprise actually appear to have improved since the buyout, recording a post-tax surplus of US$0.53 million in KDHP's first year of operation (Pillai 2006). Nevertheless, the point remains that whereas under Tata ownership Munnar's tea operations could be continued despite annual losses, such opportunities do not exist with local ownership.

Estate Abandonment

The district in South India worst affected by the tea crisis is undoubtedly Central Travancore, in the Periyar region of Idukki District, Kerala. In early 2006, 19 of the 39 estates in Central Travancore were non-functioning, such that the bushes were either being plucked by workers and sold (often through trade unions) to local BLFs, or were completely abandoned and non-productive. Disaffected ex-workers, many still locked in legal battles over unpaid wages, provident funds and gratuities, had divided up plots of the estates amongst themselves. At the time of writing, the future of these large tracts of land remains highly uncertain. An NGO/trade union report on this crisis situation (CEC 2003) documented extreme poverty, starvation-induced deaths and suicides on the abandoned estates.

Why have the tea estates of Central Travancore responded in such an extreme way to the tea crisis? Ecological factors, including a comparatively low altitude (averaging around 900 metres) and lower, more seasonal rainfall, appear to have led to increased vulnerability to pest and disease in the district. (Incidences of red spider mite were high in February 2006.) It could be argued that the area is a marginal tea-growing district at the best of times. Due to these conditions, replanting had been neglected or delayed, and many of the estates are planted with aging, unproductive bushes. More contentiously perhaps, is the argument that labour relations are more fractious in Kerala, and the high minimum wages (relative to neighbouring states, that is) and less flexible wage setting mechanisms in the state have eroded profitability.

Furthermore, testimony from numerous informants (including unions, NGOs, and planter associations) point to an element of managerial opportunism precipitating abandonment. In the 1990s, many of the estates were acquired by investors with very little organisational experience in plantation management. At least one such company, the Ram Bahadur Thakur (RBT) Group, entered the sector from a mining background in the northern state of Bihar. Such enterprises appeared to use their estates as 'cash cows', extracting optimal returns when prices were high but without suitable re-investments, leading to the spectacular folding of all nine RBT estates in Central Travancore. In contrast, adherence to a 'planter culture' is sometimes associated with a longer term sense of land stewardship and responsibility to the dependent enclave communities.

Some industry observers cite the severe impacts of estate abandonment in Central Travancore as a forewarning of what will eventually transpire in other estate districts of South India. In particular, the plantation sector has argued that the impositions of the *Plantation Labor Act* are a cost burden that many competitor countries, and indeed other sectors of the Indian economy, do not have to bear. In the context of a global, audit-driven regime of traceability and documented adherence to social responsibility, these complaints may be compromised by the fact that the PLA provisions frequently provide precisely the kind of evidence that end-buyers increasingly demand. In the tea industry, the Ethical Tea Partnership (ETP) is emerging as a key driver for such documentation. It is pertinent, however, to remember that initiatives such as the ETP have not yet penetrated into the main markets for South Indian tea: the Middle East and Russia. Further, ethical trade initiatives are already developing certification models attuned also to the needs of the smallholder sector, thereby offering little additional competitive advantage to estates.

Importantly, as a result of PLA requirements, government agencies do not always provide standard services to the communities living in the plantation enclaves. This explains the severe hardship faced by workers on the abandoned estates, some of whom have been living on the estates for generations and effectively 'know no other home'. This raises the question of cost efficiencies on the estates vis-à-vis smallholders, who do not bear the same social responsibilities to their workers.

Whereas estates in South India were once reluctant to process leaf in their factories from smallholders due to quality concerns, many estates in Wyanad, the Nilgiris and elsewhere are increasingly implementing out-grower schemes in an attempt to increase throughput to maximise factory capacity.

Conclusion

The discussion in the section above identifies a wide array of restructuring pathways being executed within South Indian tea estates. Lower prices for tea since the late 1990s have reverberated across the entire sector, but have produced impacts of varying magnitudes and inspired different responses from the estates. Industry structures in South India are responding to a global private-regulation of the tea trade and the ever-increasingly importance of brand management. Two key insights provide the major findings from the chapter.

First, evidence from the South Indian tea estate sector indicates a need to temper generalised accounts of change that posit the advantaging of larger-sized entities. Tea estates in South India have not escaped the burdens of coping with the competitive pressures associated with globalisation. Indeed, as evidenced through the instances of estate abandonment, employee buyouts, and increased outsourcing to small growers, there are newly emergent tendencies for large parent companies to exit the industry leading to the re-localisation of ownership structures. Secondly, these tendencies are not yet crystallising into a single dominant form. Invariably it is tempting to compress such variability into an account that identifies an ascendant meta-narrative of change. To apply such an approach to the current case, however, would be ingenuous. Rather than 'writing out' variability for the sake of a coherent narrative that sees current restructuring processes converging ultimately towards a particular direction, we contend that themes of variability, difference and experimentation very much need to be at the epicentre of developing an understanding of this industry.

Taking this insight as a cue, therefore, we suggest that an appropriate agenda for value chain analysis is to explore in greater detail the diversity of strategic responses to contemporary global economic pressures, and how the institutional environments of regional production centres interact with global value chains to produce highly varied responses. Outcomes are not pre-ordained, but that the playing out of strategies, experiments and struggles configures how producers are inserted within supply chains and, more to the point, the economic returns and level of control they can exercise within them.

Acknowledgements

This chapter is based on research performed as part of an Australian Research Council (ARC) Discovery Grant 'Traceability and Developing Countries Agriculture'. The authors would like to acknowledge the support of the ARC and also that of the United Planters Association of Southern India (UPASI) in Coonoor, who facilitated field work activities in India.

References

Baffes, J. (2005), 'Reforming Tanzania's tea sector: a story of success?', *Development Southern Africa*, vol. 22, pp. 589–604.

Barrett, H.R., Browne, A.W. and Ilbery, B.W. (2004), 'From farm to supermarket. The trade in fresh horticultural produce from Sub-Saharan Africa to the United Kingdom', in Hughes, A. and Reimer, S. (eds), *Geographies of Commodity Chains*, Routledge, London, pp. 19–38.

Birthal, P.S., Joshi, P.K. and Gulati, A. (2005), 'Vertical coordination in high-value commodities: Implications for smallholders', *Discussion Paper 85*, Markets, Trade and Development Division of International Food Policy Research Institute (IFPRI), Washington D.C.

Busch, L. and Bain, C. (2004), 'New! Improved? The transformation of the global agrifood system', *Rural Sociology*, vol. 69, pp. 321–46.

CEC (2003), Crisis in Indian tea industry: A Report of the Fact Finding Team that Visited the Tea Plantations of Kerala and Tamil Nadu, Centre for Education and Communication (CEC), New Delhi.

Daviron, B. and Ponte, S. (2005), *The Coffee Paradox: Global Markets, Commodity Trade and the Elusive Promise of Development*, Zed Books, London.

Dolan, C. and Humphrey, J. (2000), 'Governance and trade in fresh vegetables: the impact of UK supermarkets on the African horticulture industry', *Journal of Development Studies*, vol. 37, pp. 147–76.

FAO (2006), Current Market Situation and Medium Term Outlook, Intergovernmental Group on Tea Seventeenth Session, Nairobi, Kenya (29 November – 1 December, 2006), Intergovernmental Group on Tea, Document No: CCP:TE 06/2, Food and Agriculture Organisation of the United Nations, Rome.

Fold, N. (2001), 'Restructuring of the European chocolate industry and its impact on cocoa production in West Africa', *Journal of Economic Geography*, vol. 1, pp. 405–20.

Friedberg, S. (2003), 'Cleaning up down south: supermarkets, ethical trade and African horticulture', *Social and Cultural Geography*, vol. 4, pp. 27–43.

Gibbon, P. and Ponte, S. (2005), *Trading Down: Africa, Value Chains, and the Global Economy*, Temple University Press, Philadelphia.

Gilbert, C.L. (1996), 'International commodity agreements: an obituary notice', *World Development*, vol. 24, pp. 1–19.

Goodman, D. (2003), 'The quality "turn" and alternative food practices: reflections and agenda', *Journal of Rural Studies*, vol. 19, pp. 1–7.

Hu, D., Reardon, T., Rozelle, S., Timmer, P. and Wang, H. (2004), 'The emergence of supermarkets with Chinese characteristics: challenges and opportunities for China's agricultural development', *Development Policy Review*, vol. 22, pp. 557–86.

Hughes, A. (2005), 'Responsible retailers: ethical trade and the strategic re-regulation of cross-continental food supply chains', in Fold, N. and Pritchard, B. (eds) *Cross-Continental Food Chains*, Routledge, London, pp. 141–54.

Humphrey, J. (2006), 'Horticulture: responding to the challenges of poverty reduction and global competition', in Batt, P. (ed.) *Acta Horticulturae, 699, Proceedings of the 1st International Symposium on Improving the Performance of Supply Chains in the Transitional Economies*, pp. 19–38.

Neilson, J. (2005), 'The politics of place: geographical identities along the coffee supply chain from Toraja to Tokyo', in Fold, N. and Pritchard, B. (eds) *Cross-continental Food Chains*, Routledge, London, pp. 193–206.

Neilson, J. (2007), 'Institutions, the governance of quality and on-farm value retention for Indonesian specialty coffee', *Singapore Journal of Tropical Geography*, vol. 28, pp. 188–204.

Neilson, J. and Pritchard, B. (2006), 'Traceability, supply chains and smallholders: case-studies from India and Indonesia', *Proceedings of the Seventeenth Session of the Intergovernmental Group on Tea in Nairobi, Kenya, November 29 – December 1*, Document No: CCP:TE 06/4, Food and Agriculture Organisation of the United Nations, Rome.

Neilson, J. and Pritchard, B. (2007), 'Green coffee? Contestations over global sustainability initiatives in the Indian coffee industry', *Development Policy Review*, vol. 25 no. 3, pp. 311–31.

Neilson, J. and Pritchard, B. (forthcoming), *Value Chain Struggles*, Blackwell, Oxford.

Neilson, J., Pritchard, B. and Spriggs, J. (2006), 'Implementing quality and traceability initiatives among smallholder tea producers in Southern India', *International Society of Horticultural Science (ISHS) ACTA Horticulturae 699: International Symposium on Improving the Performance of Supply Chains in Transitional Economies*, Batt, P. (ed.) pp. 327–34.

Pillai, R.R. (2006) '"Employee buyout" model in tea plantations tastes success', *The Hindu Online*, 18 December 2006.

Ponte, S. (2002), 'The "latte revolution"? Regulation, markets and consumption in the global coffee chain', *World Development*, vol. 30, pp. 1099–1122.

Pritchard, B. (2000), 'The tangible and intangible spaces of agro-food capital', Symposium on Agriculture, Nature and Social Change, X World Congress of Rural Sociology, Rio de Janeiro, Brazil (unpublished, available from author).

Reardon, T., Codron, J.-M., Busch, L., Bingen, J. and Harris, C. (2001), 'Global change in agrifood grades and standards: agribusiness strategic responses in developing countries', *International Food and Agribusiness Management Review*, vol. 2, pp. 195–205.

Reardon, T. and Timmer, P.C. (2007), 'Transformation of markets for agricultural output in developing countries since 1950: How has thinking changed?', in Evenson, R.E., Pingali, P. and Schultz, T.P. (eds), *Volume 3 Handbook of Agricultural Economics: Agricultural Development: Farmers, Farm Production and Farm Markets*, Elsevier, Amsterdam.

Sanjith, R (2004), 'Tea prices: north-south divergence', *Planters' Chronicle*, June, pp. 24–9.

Talbot, J. (2002), 'Tropical commodity chains, forward integration strategies and international inequality: coffee, cocoa and tea', *Review of International Political Economy*, vol. 9, pp. 701–734.

Tea Board of India (2003), *Tea Digest 2002*, Tea Board of India: Kolkata.

Tea Board of India (2005), *Tea Digest 2004*, Tea Board of India: Kolkata.

Weatherspoon, D. and Reardon, T. (2003), 'The rise of supermarkets in Africa: implications for agrifood systems and the rural poor', *Development Policy Review*, vol. 22, pp. 333–55.

Chapter 5

Contesting the Rules of Exchange: Changing Conventions of Procurement in the MERCOSUR Yerba Mate Commodity Chain

Christopher Rosin

Introduction

The concept of the commodity chain is a frequently utilised tool in the analysis of agricultural production systems. Despite its potential to impose hierarchical structures or to ignore relations skipping links in the 'chain' (Raikes *et al.* 2000; Smith *et al.* 2002), the approach offers substantial insight to the incorporation of agricultural products into commodity markets (Jackson *et al.* 2006). Notably, commodity chain analysis delineates a readily distinguishable whole within the complexity of the agricultural sector. As such, it facilitates a concerted focus on the relations and the context of those relations between participants in a designated chain (Marsden *et al.* 2000; Cook *et al.* 2004). This contribution is particularly apparent in analyses that examine and generalise production at a global scale (Ponte and Gibbon 2005).

Some limitations are, however, evident in typical applications of the commodity chain concept within a political economy perspective. Foremost among these is the tendency for a given chain to appear as an entrenched structure subject only to shifts in the features of and processes associated with capitalist penetration of agricultural markets. This aspect of commodity chain analysis enhances its focus on the existing political economic structures which help to determine the parameters of production for any given commodity. Thus, for example, comparing a commodity chain structured along Fordist principles with one of a more flexible nature can illuminate pertinent influences of the global economy. Analysis of relations within such commodity chains also offers some indication of the political economic factors that define the relative capacity of actors within each chain to affect issues of social justice, distribution of value, and sustainability. Despite these potential insights, the delineation of a commodity chain fails to account for the processes through which its structures are established and continue to develop.

The relative benefits and limitations of a commodity chain framework are associated with political economic approaches to agri-food systems more generally. A common criticism involves the focus on class, power and social relations to the exclusion of the role of objects, nature, and non-human beings. Actor Network

Theory (ANT) has been adopted by some researchers for whom the concept of networks offers a better means to develop symmetrical relations between humans and things (see Goodman 1999; Murdoch 2000). Murdoch (2001, 129), however, acknowledges the potential drawbacks of the ANT approach which, in practice, often fails to sufficiently acknowledge the 'exemptionalism' of humans – that is, their capacity to enact change on environments and to hold ethical positions.

An incisive analysis of contemporary agri-food systems must involve the recognition of the multiple and complex relations involved in the process of getting food from the farm to the plate. How can the advantages of the commodity chain concept be exploited while examining the development of exchange relations that contribute to the distinctive temporal manifestations of a particular commodity chain? Is it possible to approach these relations and their emergence as the purposive actions of humans who incorporate non-human beings and objects as active elements of social interaction? By examining these alterations, can the processes through which power differentials become embedded and reinforced while remaining contested be exposed?

In this chapter, I address these questions by applying a particular theoretical tool – convention theory (CT) – to a specific context – the implementation of the MERCOSUR customs union in South America. In doing so, I demonstrate that a modification of the macroeconomic context of a commodity chain initiates changes in relations between actors at the supply end of the chain. CT brings both a more pragmatic and active perspective to commodity chain analysis. The approach involves an examination of the conventions (that is, the understandings that are formalised in rules or norms and those that simply remain agreements regarding the expected frame of action) that facilitate social interaction. Beyond merely theorising the existence of conventions, however, CT suggests that actors engage conventions strategically. Thus, commodity chain participants negotiate conventions in order to promote their relative economic position or to mitigate emerging uncertainties.

The MERCOSUR regional free trade agreement reflects the trend toward neoliberal influences on global economies (Liverman and Vilas 2006; Peck and Tickell 2002). As such, it is a prime example of the context in which the conventions of exchange experience significant change. In 1995, Argentina, Brazil, Paraguay and Uruguay began removing barriers on trade between member countries. Beyond creating a larger market, this action dramatically altered competition from a domestic context with similar constraints on production to an international one involving competitors with unknown characteristics. The resulting uncertainties encouraged the exploration of alternative means to coordinate production in the emerging economic environment. The alteration of conventions involves a negotiation process in which actors (employing distinct economic, political and knowledge power) promote changes that meet more or less receptive responses from other participants in the chain. These factors are especially relevant in the yerba mate (a tea that is a part of everyday consumption in MERCOSUR countries) commodity chain.

This analysis begins with a brief overview of CT, noting its application within agri-food analysis while distinguishing my approach from recent applications to agricultural and rural production. I then illustrate my argument using case studies of yerba mate production in Brazil and Paraguay. This section begins with a presentation of the industry in the MERCOSUR region, identifying its particular relevance for the study undertaken. Then, producer response (as recorded in a series of semi-structured interviews with 51 small-scale farmers in Brazil and Paraguay between October 1998 and May 1999) to new terms of exchange related to the procurement of raw material in the commodity chain are examined. This examination demonstrates the variable success of the changes as well as identifying the potential means by which new conventions are contested. I conclude that uncertainties associated with MERCOSUR appear, at least immediately following its implementation, to advantage the position of processors in their relations with producers in the yerba mate commodity chain.

Conventions Theory

The emergence of a French school of Convention Theory is generally attributed to Boltanski and Thévenot's (2006[1991]; 1999) examination of justifications for challenging established social practice. This work defines six worlds of justification based in distinct orders of qualification: market (prices and wealth); industrial (efficiencies and standards); domestic (status); civic (common good); fame (public affirmation); and inspired (commitment to higher goals). The relation of justifications to conventions is twofold. First, a given world of justification can define mutually acceptable frameworks for mounting challenges to undesirable situations, although compromises are often necessary to bridge contradictions between justifications. Second, where challenges to existing relations are mounted, actors strategically engage that world of justification which most benefits their position and argument. Thus, Boltanski and Thévenot claim that conventions develop in order to facilitate social interaction in situations where various justifications (and the objects and tests associated with these) are brought to bear. As such, conventions can range from localised unwritten agreements to codified international law. Critical moments – when existing conventions fail to bridge the uncertainties in social interaction – potentially give rise to strategic renegotiation among actors (Thévenot 2002). This focus on the processes through which social interaction evolves, solidifies and redevelops is ideally suited for examining economic relations within commodity chains, especially as these are affected by changes in economic structures.

CT, as developed by Boltanski and Thévenot, has occasionally contributed to the study of agri-food systems. Often, justifications are associated with the qualities of a commodity (Busch 2000). For example, CT has been used to examine aspects of fair trade (e.g., Raynolds 2004; Renard 2003) and geographical labelling (e.g., Barham 2003; Murdoch *et al.* 2000). In this literature, civic or domestic justification – as a feature of fair trade and geographically labelled agricultural goods – is contrasted with the market justification commonly associated with conventional food networks. Thus, this work emphasises the comparison of alternative and conventional markets as established entities based on the perceived justification employed by the consumer (or promoted by the seller). The literature does not, however, assess the processes through which these relations develop within the commodity chain.

Insights from a CT approach are also more specifically applied to a commodity chain framework in the agricultural sector. For example, Marsden, Banks and Bristow

(2000) argue that focusing on the conventions underlying quality designations contributes substantial insight to the emergence of shorter commodity chains as a strategic marketing tool for farmers. Ponte and Gibbon (2005) incorporate worlds of justification (albeit limiting consideration to market, industrial, and domestic worlds) to global value chain analysis. The contrast between value chains dominated by these alternative worlds is associated with the concept of buyer and producer driven chains. As with the literature on qualities, however, this work examines the resulting strategic advantage associated with competing justifications without considering the processes through which these become accepted and embedded in given commodity chains.

As a viable means of delineating the processes through which social relations develop, CT provides an excellent means for assessing the response of the actors within a given commodity chain to the impact of changing economic environments. In the approach followed here, CT is utilised as a tool to examine the 'economic and the social [...] combined in the world of things, and the procedures of qualifying the latter [as] so many occasions for stakes, for quarrels between varied and contradictory logics' (Dosse 1999, 109–110). More specifically, I employ a CT perspective to assess the response of small-scale yerba mate producers to the implementation of regional free trade. First, the perceived competitive threats associated with MERCOSUR have encouraged processors to impose tests from an industrial world of justification. Where these tests – ranging from quality standards to harvest practices – challenge existing means of establishing the worth (or relative quality) of raw material in the commodity chains, they elicit challenges from producers that are founded in a variety of worlds of justification.

Producer Response

Yerba mate is a tea produced from the leaves of a holly tree (*Ilex paraguariensis*) endemic to the forests of the MERCOSUR countries. Because of its significant role in local diets and cultures, production of the tea is considered a vital element of food security for Argentina, Brazil and Paraguay. Prior to 1995, tariffs on yerba mate enforced by the governments of these countries established protected and stable domestic markets. Processors generally developed sales regions reflecting their ability to cultivate relationships with individual retailers. Access to raw material was, however, limited by difficulties associated with its cultivation. In this situation, conventions emerged that allowed processors to secure raw material while assuring the sale of product for producers. The creation of MERCOSUR significantly altered the production and marketing of yerba mate by introducing free trade between the member countries.

Given the strong domestic orientation of the respective yerba mate industries, the implementation of MERCOSUR created two sources of uncertainty in the Brazilian and Paraguayan commodity chains. The greater capital capacity of Argentine firms threatened processors in the former countries with eventual competition with final product imported from Argentina. A more immediate concern was the introduction of a new (imported) source of raw material upsetting existing market patterns. Processors in Brazil and Paraguay responded to these threats both individually and collectively. (Rosin (2007) presents a more detailed examination of the processors' actions.) Among the latter actions, processors attempted to reconstruct the conventions of raw material procurement by invoking industrial justifications in their exchange relations with producers. In Brazil this involved attempts to shift the responsibility for the harvest from gangs hired by the processors to the individual producers. Brazilian processors also encouraged producers to cultivate more intensive plantations that could be harvested throughout the year in order to facilitate continuous plant operation and reduce storage needs. In Paraguay the emphasis was on raw material quality including freshness (24-hour delivery window) and the size of harvested branches, standards which increased producers' exposure to risk and reduced the mass of harvested material. By attempting to reduce the direct and indirect costs of securing raw material, processors sought to shift the impact of MERCOSUR competition toward the producers.

Whereas yerba mate processors either understood the implications of MERCOSUR or were associated with people who did, few of the producers had achieved similar levels of familiarity with the new market. As a result, producers' responses to attempts to restructure exchange relationships was framed in the context of past relations with processors and their experience with yerba mate as a commodity. This situation contributed to uncertainties within the respective commodity chains as the industrial justifications proposed by processors undermined the domestic and market justifications that had stabilised traditional exchange relations. In the following section, data from producer interviews demonstrate their responses to the uncertainties associated with an altered justification of exchange relations.

Examination of producer response in four study sites (Brazil's Chapecó and Canoinhas and Paraguay's Natalio and Colonia Independencia regions, see Figure 5.1) indicates the extent to which past contexts of exchange relations influence producers' capacities to negotiate conventions aimed to address industrial justifications. Each site is characterised by distinct producer-processor relations and varying levels of experience with industrial justifications in exchange relations both of which affect the producers' capacity to negotiate the terms of emerging conventions. As a result of these differences, only in Colonia Independencia has a growing emphasis on efficient procurement of quality raw material initiated significant alterations of local production systems. A short overview of the first three sites demonstrates the lack of strong contradictions in the justifications experienced there. Colonia Independencia, where industrial justifications directly conflict with existing domestic justifications of exchange, will be discussed in more detail. The initial conclusions drawn from this analysis reflect the instability of the yerba mate commodity chain at the time of the interviews.

Chapecó, Canoinhas, and Natalio

In the Chapecó, Canoinhas, and Natalio study sites, MERCOSUR's implementation had minimal impact on the yerba mate commodity chain from the producers' perspective. As a result, the processors' credibility is not challenged by producers responding to the industrial tests the former attempted to impose. By contrast, the

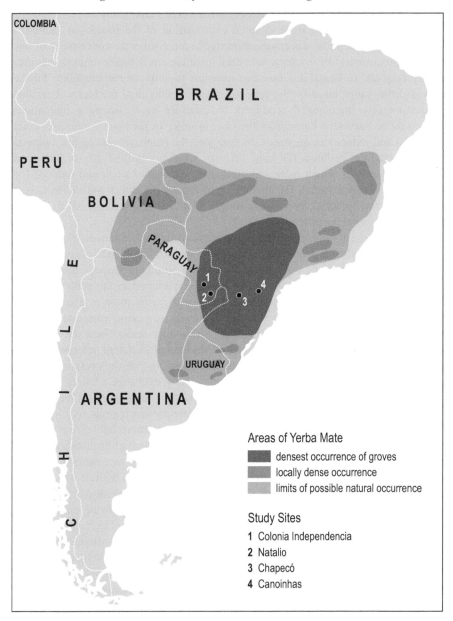

Figure 5.1   Map of study sites relative to natural distribution of yerba mate

Source: Adapted from Aranda n.d.

producers are more prone to assess the relative worth of yerba mate as a cash crop based on perceptions of its labour inputs, perceived returns, and current price in comparison to alternative crops (i.e., market justifications). The limited influence of the processors' demands on the position of the producers within the commodity chain can be attributed to the existing exchange relations and the relative importance of the crop.

The relationship between producers and processors in Chapecó, Canoinhas, and Natalio is mediated by third parties who are responsible for the procurement of the raw material. In the Brazilian sites this situation reflects existing conventions that were defined, in part, by the producers' limited dependence on yerba mate income. To ensure a more regular supply and avoid competition for labour from other crops, the processors hired harvest crews who travelled up to 250 km to procure raw material. Despite these advantages, the costs of employing the crews were considered unsustainable under existing market conditions. Faced with competition from low-cost imported yerba mate, many processors relying exclusively on domestic product offered incentives (equivalent to US$0.05/kg) for raw material delivered directly to the firm. At the time of the interviews, few producers appeared willing to devote their time and labour to earn these incentives. In the Natalio site, by contrast, the intermediaries performed a necessary service in the commodity chain by providing the means to transport the raw material to the processor. As in the rest of Paraguay, the harvest is concentrated during June, July and August. In comparison to Brazil, the timing of harvest remains a principal factor associated with the quality of the raw material. Because processors in Paraguay sell a product that has aged at least two years and can be taken from storage at any time of year, they are not as dependent on a consistent, year-round supply. As such, the intermediaries acted on their own initiative – engaging in the yerba mate commodity chain when prices ensured a profit for the truck owner. In all these sites, the intermediaries faced the quality standards established by the processors. The industrial justification underlying the standards was subsequently translated into a market justification based on the price the intermediaries offered and at which the producers were willing to sell.

The worth of yerba mate as a cash crop was also defined by justifications superseding those introduced by the processors. For example, in Chapecó, yerba mate has only become a feature of farming systems over the last three decades. The crop was generally perceived as an alternative to soybeans, maize, and blackbeans that could be cultivated in sites inaccessible to farm machinery. Plantations occupied relatively small areas (0.25 ha) and contributed marginally to farm incomes. Several farmers did not cultivate yerba mate specifically because it required labour not accomplished with a tractor. Most Chapecó producers complained about low prices and some indicated that they were not paid in a timely manner. Few, however, were aware of the increasing quality demands of the processors.

In Canoinhas, yerba mate owns a reputation as a failed crop – thereby claiming limited worth in a justification based on fame. Until the 1950s, it was the region's primary cash crop benefiting from abundant naturally-occurring trees. Many farmers recalled the bountiful harvests and greater profits of the past.

> Soon it will disappear. Because here … I remember when this was all yerba mate. And if I cut eleven trees it gave 20–25 loads of yerba mate, the amount that would fit in the wagon … But it went, it went, it went … The trees began to dry up.

Canoinhas, 12/01/99

The profitability of the crop also declined as processing firms assumed greater control over the commodity chain. Historically, producers completed the initial processing and were reimbursed for the value added. They are now paid for leaf 'on the tree' and only sell when a harvest crew is nearby. The perceived failure of yerba mate is intensified by the relative success of tobacco in the region. Farmers in Natalio more consistently employ market justifications in assessing the relative worth of yerba mate, assessing the costs associated with production relative to the expected returns. As perennial crops with relatively low maintenance requirements yerba mate or tung (an oil crop) were viewed as potential alternatives to soybeans. Tung was preferred by many because, as the following farmer argues, it promised better returns:

> I was thinking [...] about which crops I have and why. I have tung because there are few costs – just a little cleaning [*limpiesa*] – and it gets a good price ... Oh, and there is the yerba mate, but it has a very low price now. It barely pays to keep the plantation clean.

Natalio, 14/04/99

For those planting yerba mate, the practice often reflected a commitment to mixed agroforestry systems, something for which the larger tung trees were not appropriate. Producers in the Natalio site frequently complained about the price of yerba mate, but did not challenge industrial tests on raw material procurement.

In these study sites, the imposition of tests derived from industrial justifications had little impact on the production end of the yerba mate commodity chain. Because the producers were not directly exposed to the demands regarding product quality and efficiency of delivery, they showed little interest in negotiating new conventions of exchange beyond their continued participation in the commodity chain. In addition, the potential for industrial justifications to establish the relative worth of yerba mate was obscured by justifications attached to other, more dominant cash crops in each site. In comparison, strong patron-client relationships and the importance of yerba mate as a source of income in Colonia Independencia contributed to greater contestation of processors' actions in that site.

Colonia Independencia

The final study site, located near Colonia Independencia in central Paraguay, provides a more vivid example of the potential contradictions associated with the promotion of new forms of justification. In the past, relative isolation and historical reliance on yerba mate as a commodity contributed to the development of strong patron-client relations based on domestic justifications. As a result, producers and processors in the region engaged in more personal interaction. These existing conditions heightened the impact of an increased emphasis on industrial justifications in the procurement of raw material.

Prior to the implementation of MERCOSUR, the stability of the patron-client relations began to weaken. A weak Paraguayan cotton sector (largely attributed to the arrival of the cotton boll weevil) initiated a severe stagnation of the rural economy. The impact of the economic decline was such that consumption of yerba mate decreased significantly, leaving processors with large surpluses and little ability to raise prices for the final product. As a result, prices for raw material declined, its sale no longer guaranteed and the processors less willing to extend credit to producers. The perceived threat to the Paraguayan industry associated with MERCOSUR further intensified uncertainties within the commodity chain, which processors sought to mediate through the introduction of purchasing contracts.

The response of local producers to the imposition of industrial justifications on raw material procurement ranged from open rejection to passive acceptance of the processors' actions. Criticism of the processors and abandonment of the sector were common. The former response often invoked the domestic justifications that had defined the patron-client system and referred to proposed industrial tests (quality and efficiency standards) as the betrayal of established relations. The following producer, for example, suggested that the provision of credit and a more certain market were the processors' responsibility:

> Yerba mate does not pay any more. Our *patron* does not come here now. He used to come and help us with credit to buy food and other things and we would pay him back after the harvest. But now, we can not even sell our yerba mate.

> Colonia Independencia, 01/06/99

Other farmers challenged the industrial tests as providing benefits primarily for the processors. Among like-minded producers, it was common to neglect the maintenance of plantations. The capacity of producers to abandon yerba mate was limited, however, by the viability of alternative cash crops such as sugar cane, cotton, or soybeans.

Producer response to the imposition of quality and efficiency standards included some more positive perspectives. A minority of producers in the study area continued to view the crop as an important element of their farms' production. While not fully embracing the ascendancy of an industrial order of justification, such producers (generally those with a reputation for high quality and/or access to larger quantities) developed means to maintain and improve their relationships with a particular processor. From their perspective, continued participation in the sector involved other justifications. For example, one family was rewarded – or achieved a higher worth – due to the prestige (fame) of meeting standards:

> The [processor who] buys our yerba mate [...] made us sign a contract that we would only bring smaller stems and fresh leaf. The first time he checked our leaf, but now he has confidence that we bring good quality, so he takes ours without looking. He does not buy from some people though ...

> Colonia Independencia, 03/06/99

Other farmers suggest that continued yerba mate production was more a reflection of the relationship with a particular crop as opposed to any of the processors:

> I grow yerba mate because I like it. I would rather plant trees than other crops. I like to watch them from year to year and see how they grow, how I care for them. I add a few more trees when I have time. I know that every year I will get something from them.

> Colonia Independencia, 02/06/99

Among such farmers, the preference to 'care for' (*cuidar*) the tree and take pride in the presentation of their plantation demonstrated reference to an inspired world of justification. Their worth was established by attaining a higher objective of being a 'good' farmer and pursuing an enjoyable profession.

A CT approach to the emerging conditions of the production and exchange in the yerba mate commodity chain in Colonia Independencia facilitates explanation of the contested and contingent nature of its (re)construction. By focusing on the justifications employed by the producers confronted with new tests based in the industrial world of justification, it is possible to develop an explanation of the emerging nature of exchange relations following the implementation of MERCOSUR. The actions pursued by many of the producers challenged the processors' power to implement these tests by withdrawing a source of supply. It is also possible, however, to identify several compromises employed by producers engaging fame or inspired orders of justification.

Conclusion

It is evident that MERCOSUR initiated fundamental changes in the social relations circumscribing the Brazilian and Paraguayan yerba mate commodity chains. Both more extensive competition and reinforcement of free market concepts challenged existing conventions of raw material procurement. In response to these pressures, processors in both countries attempted to introduce practices to improve the efficiency of the industry and the quality of its output. By interpreting the processors' actions as the implementation of tests based on an industrial world of justification, a CT perspective enables an incisive analysis of the processes of reconstruction in the commodity chain.

The increased emphasis on industrial justifications to define the exchange of yerba mate leaf unevenly affected small-scale producers in Brazil and Paraguay. In regions characterised by less personalised relations with processors, intermediaries responsible for harvest have assumed the brunt of the impact. To the extent that the processors' actions affected the costs of harvesting, intermediaries translated industrial justifications into market terms by offering higher prices for more easily harvested yerba mate. In Canoinhas, Chapecó and Natalio, small-scale producers generally interpreted the situation as a further indication of the limited economic viability of the crop.

A distinct situation defines Colonia Independencia where the introduced industrial tests more directly contradict existing patron-client relations. As a result, the commodity chain experienced significant pressure to shift to a stable constellation of substantially different conventions. Producer response to the emerging conventions suggests that MERCOSUR will be the catalyst for a spatial reorganisation of yerba mate production, with the crop more likely located outside the catchment of the local sugar industry. What remains of yerba mate production in Colonia Independencia will involve individuals who reconcile the contradictions of the industrial and domestic justifications by employing tests from inspired, fame or other worlds of justification. While it is evident in this situation that the processors exerted greater economic and social power, their attempts to restructure the commodity chain and reduce their exposure to the risks and costs of competition within the MERCOSUR market met with only partial success.

These findings attest to the potential value of applying CT within a commodity chain framework. By exposing the diverse actions (and justifications of those actions) of participants at a critical moment of contradiction, the approach enables explanation of the development of stable constellations of social relationships that underpin commodity chains. Furthermore it highlights the emergent properties of commodity production and exchange, acknowledging the iterative processes of challenge and contestation of exchange relations based in the range of justifications to which the participating actors (including retailers, consumers, and others who were not part of this analysis) refer. Finally, it avoids the potential limitations of an ANT approach by locating the purposive actions of producers and processors within a world inhabited by things such as yerba mate trees, rates of deterioration of harvested leaves, sugar cane, and production contracts all of which influence the features of the emerging yerba mate commodity chain.

Acknowledgements

Christopher Rosin gratefully acknowledges the collaboration of EPAGRI staff (in both Chapecó and Canoinhas, Brazil) in the research on which this chapter is based. They provided invaluable assistance in developing contacts with both the processors and producers of yerba mate in their respective regions.

References

Barham, E. (2003), 'Translating *terroir*: the global challenge of French AOC labeling', *Journal of Rural Studies*, vol. 19, pp. 127–38.

Boltanski, L., and Thévenot, L. (1999), 'The sociology of critical capacity', *European Journal of Social Theory*, vol. 2, pp. 359–77.

Boltanski, L., and Thévenot, L. (2006)[1991], *On Justification: Economies of Worth*, Princeton University Press, Princeton.

Busch, L. (2000), 'The moral economy of grades and standards', *Journal of Rural Studies*, vol. 6, pp. 273–83.

Cook, I. *et al.* (2004), 'Follow the thing: papaya', *Antipode*, vol. 36, pp. 642–64.

Dosse, F. (1999), *Empire of Meaning: The Humanization of the Social Sciences*, University of Minnesota Press, Minneapolis.

Goodman, D. (1999), 'Agro-food studies in the 'age of ecology': nature, corporeality, bio-politics', *Sociologia Ruralis*, vol. 39, pp. 17–38.

Jackson, P., Ward, N. and Russell, P. (2006), 'Mobilising the commodity chain concept in the politics of food and farming', *Journal of Rural Studies*, vol. 22, pp. 129–41.

Liverman, D.M. and Vilas, S. (2006), 'Neoliberalism and the environment in Latin America', *Annual Review of Environmental Resources*, vol. 31, pp. 327–63.

Marsden, T., Banks, J. and Bristow, G. (2000), 'Food supply chain approaches: exploring their role in rural development', *Sociologia Ruralis*, vol. 40, pp. 424–38.

Murdoch, J. (2001), 'Environmental sociology and the ecological challenge: some insights from actor network theory', *Sociology*, vol. 35, pp. 111–33.

Murdoch, J. (2000), 'Networks – a new paradigm of rural development?', *Journal of Rural Studies*, vol. 16, pp. 407–419.

Murdoch, J., Marsden, T.K. and Banks, J. (2000), 'Quality, nature, and embeddedness: Some theoretical considerations in the context of the food sector', *Economic Geography*, vol. 76, pp. 107–125.

Peck, J. and Tickell, A. (2002), 'Neoliberalizing space', *Antipode*, vol. 34, pp. 380–404.

Ponte, S. and Gibbon, P. (2005), 'Quality standards, conventions and the governance of global value chains', *Economy and Society*, vol. 34, pp.1–31.

Raikes, P., Jensen, M.F. and Ponte, S. (2000), 'Global commodity chain analysis and the French filière approach: comparison and critique', *Economy and Society*, vol. 29, pp. 390–417.

Raynolds, L.T. (2004), 'The globalization of organic agro-food networks', *World Development*, vol. 32, pp. 725–43.

Renard, M.–C. (2003), 'Fair trade: quality, market and conventions', *Journal of Rural Studies*, vol. 19, pp. 87–96.

Rosin, C. (2007), 'Justifying the "alternative": renegotiating conventions in the yerba mate network', in Maye, D., Holloway, L. and Kneafsey, M. (eds), *Constructing 'Alternative' Food Geographies: Representation and Practice*, Elsevier, Amsterdam, pp. 115–132.

Smith, A., Rainnie, A., Dunford, M., Hardy, J., Hudson, R. and Sadler, D. (2002), 'Networks of value, commodities and regions: reworking divisions of labour in macro-regional economies', *Progress in Human Geography*, vol. 26, pp. 41–63.

Thévenot, L. (2002), 'Conventions of co-ordination and the framing of uncertainty', in E. Fullbrook (ed.), *Intersubjectivity in Economics: Agents and Structures*, Routledge, London, pp. 181–97.

Chapter 6

Audit Me This! Kiwifruit Producer Uptake of the EurepGAP Audit System in New Zealand

Christopher Rosin, Hugh Campbell and Lesley Hunt

Introduction: Audit, Neo-liberalism and Sustainability

Audit systems are common features of global commodity chains, providing a means to assure consumers of the quality standards and production practices of products sourced from unfamiliar people and places. As such, they are increasingly recognised as important elements of commodity chain governance in a context of neo-liberalism and globalisation (see Larner and Le Heron 2002; Shore and Wright 1999). Scholars in the agri-food tradition frequently claim that the development of sophisticated and powerful technologies of audit – often embedded within sites of complex negotiation between the diverse participants in these systems – is one of the most interesting features of contemporary global agri-food systems. The emergence of this interest is signalled in two benchmark papers. Le Heron (2003) establishes the central role of systems of audit, grading and standards in the emergence of specifically neo-liberal forms of food governance. In a similar vein, Busch and Bain (2004) identify the proliferation of grades and standards as the ironic unintended consequence of neo-liberal deregulation of global food trading. Since the publication of these two papers food auditing has entered debates around neo-liberalism, new forms of governance, and the relationship of these to globalising commodity systems.

A significant body of literature has also emerged that examines the potential relationship between audit standards and the promotion of a more sustainable agriculture. Precursors to this work are found in debates about the style and outcomes of a particular, commercialised, development pathway for organic agriculture. Buck et al. (1997) situated organic agriculture more as a new entrant into the dynamics of mainstream agri-food commerce than as an alternative model of production. These claims initiated a debate regarding the best way to interpret the structural implications of 'commercial organic' in the world food economy (see, for example, Coombes and Campbell 1998; Guthman 1998; Hall and Mogyorody 2001). While the main discussion focused on the *conventionalisation* or the *structural bifurcation* of organic production, the dynamic consequences of formal certification and inspection as a new mode of governance over food was also identified. For example, Campbell and Liepins (2001) examined the specific politics of increasingly formalised organic inspection in New Zealand revealing important mechanisms to preserve claims of production system sustainability associated with audits.

The arrival of a powerful new European-based alternative to organic certification organised around an alliance of large food retailers further informed analysis of audit systems for food production. Increasing interest in this new alliance – EurepGAP – and its influence on global-scale agri-food supply chains into Europe has developed in Australia and New Zealand. This literature identifies EurepGAP as a new form of agri-food governance (Campbell 2005), examines its relationship to neo-liberal reform of agriculture and to the declining influence of organic agriculture as the main vehicle for advancing commercial claims towards agricultural sustainability (Campbell *et al.* 2006a; Campbell *et al.* 2006b), and places it within a wider shift in power relations between the production and consumption ends of the agri-food system (Campbell and Le Heron 2007).

While there is strong evidence of the structural influence of EurepGAP as a new entrant into agri-food governance – especially in regard to sustainable production, the extent to which it influences the practices of complying growers remains open to debate. Guthman's (2000; 2004) examination of the sustainability claims of organic producers during the commercialisation of the industry addressed similar issues. She concluded that the commercial phase of development undermined key sustainable practices that had characterised earlier, less commercial, forms of organic production. In this chapter we apply Guthman's questions about grower-level uptake of sustainable practice to their compliance with EurepGAP standards in the New Zealand Kiwifruit industry. Data for this assessment is drawn from interviews with 36 kiwifruit growers conducted under the auspices of the ARGOS research programme.[1] Before presenting the growers' perspectives, however, we review the development of EurepGAP and the structure of the New Zealand kiwifruit sector as the context within which their response is formed.

The Emergence of the EurepGAP Audit Alliance[2]

The first attempt to retail foods using claims of sustainability in Europe involved the commercial production of organic food. Supply proved difficult to stabilise, however, and *organic* remained a market niche. Consequently, retailers sought a 'third way' to sustainable agriculture through a parallel shift of mainstream fruit and vegetable production into systems with authentic claims towards 'sustainability' (Morris and Winter 1999). They began with Integrated farming systems that promised to deliver 'residue-free' produce from systems amenable to uptake by

---

1 ARGOS is a six-year longitudinal study of 105 farms in New Zealand's kiwifruit, sheep/beef and dairy sectors. The farms are grouped geographically with equal representation of management systems relative to formally certified systems of 'sustainable' agriculture (see www.argos.org.nz). The authors acknowledge the contributions of all researchers within the project.

2 A more detailed presentation of the emergence of EurepGAP as a 'mega-audit' of food safety and sustainability in European supermarkets is presented in Campbell 2005; Campbell *et al.* 2006a; and Campbell at al. 2006b.

mainstream farmers (Howley 1997). The Integrated approach was an international science initiative to reduce pesticide usage that emphasised selected (usually 'soft') pesticides, need-based (not scheduled) applications, biological pest control, and producer monitoring.

Globally through the mid-1990s, many corporate fruit and vegetable suppliers promoted integrated production systems to source residue-free fruits and vegetables. The exact standards of integrated management, including permitted levels of chemical inputs and residue tolerance levels, were negotiated between corporate suppliers and the purchasing agents of individual supermarket chains and other multiple retailers. The resulting piecemeal proliferation of integrated production standards led to chaos in fruit and vegetable supply chains.

In 1997, a group of European retailers began discussions on integrating and standardising integrated production systems in each of their supply chains under the term, Integrated Crop Management (Howley 1997). That same year, they formally established the Euro-Retailer Produce Working Group, operating under the acronym EUREP. This group had an ambitious agenda: uniting the basic system of food safety auditing – termed Hazard Analysis and Critical Control Point (HACCP) – with a series of agricultural production practices that created a broad definition of Good Agricultural Practice (GAP). The intent was to define a series of protocols for food safety and agricultural sustainability that stretched over the whole supply chain – from farm gate to shop floor (see Table 6.1).[3]

Table 6.1   Statement of EurepGAP objectives

EurepGAP Terms of Reference
Respond to Consumer Concerns on Food Safety, Animal Welfare, Environmental Protection and Worker Welfare by:
• Encouraging adoption of commercially viable Farm Assurance Schemes, which promote the minimisation of agrochemical inputs, within Europe and world wide.
• Developing a Good Agricultural Practice (GAP) Framework for benchmarking existing Assurance Schemes and Standards including traceability.
• Providing guidance for continuous improvement and the development and understanding of best practice.
• Establishing a single, recognised framework for independent verification.
• Communicating and consulting openly with consumers and key partners, including producers, exporters and importers.

Source: www.eurep.org.

While there have been many attempts to establish food audit systems, EurepGAP is notable for its success. In the seven years since launching the standards for fruit

---

3 The result was a plethora of evolving production standards for fruit and vegetables (See, for example, EurepGAP 2001; 2004a; 2004b).

and vegetables in Europe, the majority of suppliers into the European food supply chain have been enrolled (EurepGAP Newsletter 2003–04). The current retailer membership of EUREP (noted in Table 6.2) incorporates the top corporate European supermarkets. These form an important concentration of retailer power behind the unified GAP standards for 'safe and sustainable food', significantly impacting the supply end of European fruit and vegetable commodity chains (Campbell *et al.* 2006a). In a short period, EurepGAP has become the 'gold standard' of audits for fruit and vegetable supplies into Europe and has established integrated management protocols as a serious alternative to certified organic in underwriting claims of agricultural sustainability.

Table 6.2   European Supermarket Membership of EurepGAP

ASDA/Walmart	Fedis/D.R.C.	Metro	Superquinn
Albert Heijm	ICA Hanlarna	Migros	Somerfield
COOP Italia	Kesco	Pick 'n' Pay	Tesco
COOP Norge	KF	Sainsbury	TSN
COOP Switzerland	Laurus	Safeway	Waitrose
DelHaize	Marks and Spencer	Spar Austria	
Eroski	McDonalds Europe	Superunie	

Source: Campbell et al. (2006a).

Kiwifruit Production in New Zealand[4]

One of the most successful entrants to the EurepGAP audit alliance is the New Zealand kiwifruit industry where production is dominated by ZESPRI International Ltd, the country's sole exporter. Furthermore, ZESPRI acts as the dominant corporate player in the global kiwifruit commodity chain, holding a significant market share and defining top-end quality standards. ZESPRI has been a celebrated (EurepGAP publicity freely uses ZESPRI as an example of a success story) collaborator in EurepGAP, having developed an extensive programme of compliance with EurepGAP protocols to secure market access to the elite tier of European retailers. As a result of this diligence, ZESPRI was appointed Chair of the EurepGAP technical standards committee designing the global production protocols for kiwifruit.

The alacrity with which ZESPRI embraced EurepGAP at an industry-level is firmly rooted in a significant period of crisis experienced by the kiwifruit industry in New Zealand. Facing significant price competition in the world market, particularly from new competitor regions in Chile, Italy and California, the New Zealand industry became financially insolvent in 1991. Since then, the industry has moved from a period of intense crisis to very high levels of prosperity (see Campbell *et al.* 1997; Campbell *et al.* 2006a). Multiple factors contribute to the resurgence of the industry

---

4 For a more detailed presentation of the structure of New Zealand's kiwifruit commodity chain, see Campbell *et al.* 1997.

since the late-1990s. First, a re-branding strategy was introduced (including the new brand name ZESPRI) along with promotion of a new 'gold' variety of kiwifruit. This effort was premised on the termination of bulk commodity production under intensive vine management and subsequent adoption of 'environmentally friendly' organic and Integrated management systems. This shift coincided with moves towards 'green protectionism' at a regulatory level and promotion of 'green' food consumption by supermarkets and large cooperatives in key kiwifruit markets (Campbell and Coombes 1999). In a near-revolution of kiwifruit production, a chemically intensive production sector (circa 1992–94) became 100 percent organic or Integrated by 1998, allowing ZESPRI to position itself as an elite supplier to 'green' retailers. The result was a spectacular turnaround in the industry's fortunes with consecutive record high returns to the sector from 2001–05.

When EurepGAP was launched in 1999, ZESPRI was fully aware of the bewildering array of retailer-specific audits characterising the European supply chain. ZESPRI saw EurepGAP as a significant opportunity to amalgamate the endogenously constructed environmental quality audits as well as simplify the huge number of environmental quality standards established by European retailers. By assuming the chair of the technical standards group for EurepGAP's Integrated Assurance Scheme in kiwifruit in 2003, ZESPRI was able to align its existing audit and production activities with EurepGAP, publish its own specialised version of the EurepGAP standards and institute these as the compulsory minimum standard for growers supplying the export market.

In conclusion, EurepGAP has become the mechanism by which ZESPRI and the New Zealand kiwifruit industry is moving into a privileged supply relationship relative to rival production regions (see Campbell 2005). ZESPRI's close relationship with EurepGAP is a key facet of the industry's return to prosperity. As such, the wholesale conversion to EurepGAP compliance is an important feature of the current production practices of kiwifruit producers, validating claims of sustainable production based in the application of integrated management systems. The next section addresses the grower-level ramifications of the politics of this strong alignment to EurepGAP.

Grower Responses to EurepGAP

The 36 households from which data was collected include an equal number of orchards in three panels – certified organic, green and gold – defined by certified management options for 'sustainable' kiwifruit production. Around five percent of the kiwifruit orchardists operate under certified organic management. The remaining 95 percent use integrated management systems to produce Green (*Actinidia deliciosa*) and/or Gold (*Actinidia chinensis*) kiwifruit. Since 2003, all growers are compliant with the EurepGAP audit (ZESPRI 2003). As part of the ARGOS programme, households participated in extended qualitative interviews about diverse aspects (social, economic and environmental) of kiwifruit production. More specifically, these interviews interrogated a range of perceived constraints in the management of kiwifruit orchards. Within the semi-structured interview format, all households were questioned about the auditing of sustainability including specific questions about organic and EurepGAP compliance. The following sections outline the responses of these households, demonstrating the complex way in which growers negotiate the politics of the new audit system.

Impacts on Orchard Production Practices

Because EurepGAP basically replicates the existing integrated management standards for the industry, most orchardists perceive the audit as an alteration of the politics of commodity production rather than a significant change in management practices. Substantial new processes involved the rigorous compliance audit (and associated documentation procedures) and infrastructural requirements relating to access to toilets in orchards and chemical storage. For the great majority, the audit mainly involved a shift in time allocation with much more time allocated to compliance reporting.

Negative Views of EurepGAP

The majority of growers view the actual audit imposed under EurepGAP as primarily an exercise in generating paperwork. This not only represented a change in the balance of tasks in orchard management, it also had potential impacts on their wider social life.

> Male: Socially there was probably more time available ... yeah, this paperwork, we didn't have that – so many food safety programmes that we have to fill in and EurepGAP and all the other bits and pieces. So, we had time for things outside the gate ... There is this friend of mine, he said he knocks off at the end of the day thinking, "Cripes, I have still got to fill in this paper or that paper or I've got to comply with this, I have got to comply with that, I am behind as such." So he said, "I always feel as though I am overloaded with this component of the business that I don't actually enjoy doing. It is just the paper work side of it ..." That pressure was never there. So you could go off and go to twilight cricket, or go to bowls, or go fishing, or take the kids to the beach ... whereas today you are struggling to ... (Gold)

This perception is reinforced by the fact that management practices promoted under EurepGAP differ little from existing ones except in degree or level of detail. Participants who view themselves as orchardists 'first and foremost' generally detest the audit due to the location of work (office) associated with compliance procedures. In addition to the paperwork, the excessive detail of the audit questions irritates these orchardists.

> Male: [Audit compliance is] such a complicated big machine. You know, I hate paperwork. I just like running around in the orchard. My partner sort of tends to do most of the paperwork, but I just hate it. And I probably wouldn't be organic because I couldn't stand the paperwork. The other thing I don't like is EurepGAP because you have got all that paperwork again. (Organic)

> Son: I literally, I hate it [paperwork]. I really ... I just [hate it]. Having to go through it every year.

Father: The language is difficult as well. If they used common language that we could understand. Some of the questions are almost replicated … I think in the latest format, when they came out in [ZESPRI] Road Shows last year and said [that] instead of two hundred and fifty odd questions to answer this year you have only got ninety—but the ninety, some of them had sixteen answers. (Gold)

The time required to process compliance and the increasing detail in scrutiny of orchard practice emerged as the key negative impacts of EurepGAP for growers. While EurepGAP altered little of the actual content of existing practices, it attempted to shift both the definitions of 'good' and 'bad' production practices as well as the locus from which such values claims were generated and legitimised.

Positive Perceptions of EurepGAP

While growers commonly listed the above as negative consequences of EurepGAP compliance, their responses also portrayed the audit in more positive terms. Despite criticisms of its time demands, most growers indicated some understanding of the rationale behind the EurepGAP audit. For example, the potential for an audit system to ensure food safety is rarely challenged. Furthermore, the importance of defining New Zealand kiwifruit as a premium product appears to have been accepted by the growers. Most, however, believe that current industry practices and standards already establish this quality for their fruit. Any additional value associated with the EurepGAP audit lies in its potential to control the actions of renegade orchardists.

Male: There are always loose cannons in any industry whether it is farmers, horticulturalists or whatever, there is always someone dumping waste oil down a bank, there is always someone doing something bloody silly basically, and it is just, [EurepGAP] … you know police what is happening right across the industry. That is what it is doing, it is just policing it. (Green)

For some organic orchardists, the strict detail of the EurepGAP audit potentially validates practices common in organic management. There is further speculation that, as familiarity with the process of auditing grows, some conventional growers might consider organic certification a more viable option:

Female: Maybe it's made a few conventional growers think they might as well go organic. They have got to do all this paperwork, they might as well. I think that's why a lot of them resisted [going organic], because of the paperwork. Now they've got to do it whether they're organic or not. Everybody has to do it. (Organic)

To the extent that individual growers perceived the guarantee of quality associated with EurepGAP compliance to be a priority, they are more likely to acknowledge its credibility and value. Acceptance was further enhanced for those, including former dairy farmers and teachers, having previous experience with audit systems albeit in distinct settings.

EurepGAP and Changing Political Relations in Agri-Food Systems

Whereas the preceding assessment demonstrates that growers recognise limited influence of EurepGAP on orchard management practices, it is apparent that the audit has initiated a broader reconfiguration of relations between growers and other participants in the kiwifruit commodity chain. Concerns about the increasing strictness of compliance and the number of standards have already been identified. Many growers, however, also echoed concerns voiced by those interested in sustainability more generally: that these standards are being designed and set by parties unfamiliar with the environmental and social conditions of production in New Zealand. That EurepGAP is generally associated with distant European bureaucrats who apply inflexible standards to New Zealand conditions establishes a highly contestable political terrain.

> Male: That winter chill factor is a major … constraint. And, as the climate gets warmer and warmer, it's going to get worse and worse. And now with the EurepGAP, and Europeans don't like us using chemicals, they might take Hi-Cane®[5] away from us. And if we have no Hi-Cane® then out come my vines pretty much … We know our orchard better than the Europeans, you know, and ZESPRI and down the chain. If they say on a bit of paper that I've got to do this, that and the other, I think that's bullshit … But it all depends on who's saying it and how they're saying it. You know, if they are trying to be helpful or critical.

> Son: Yeah, that came out with a lot of their questions initially, and they were totally irrelevant. Fumigation and things like this … and human sewage going out there. They were thinking in European … (Gold)

Reflecting on past experience with EU agricultural policies, some orchardists question the extent to which the EurepGAP audit works as a non-tariff barrier to market access and leads to unfair trade competition with Chile, New Zealand's main Southern Hemisphere rival.

> Male: Okay, we are asking for a premium for our fruit. And Chile, they are getting better with quality and taste and that. [ZESPRI representatives] said to me [that] Chile has come up with their own set of standards. That they have not really joined EurepGAP but, as a grower, I do not know how powerful EurepGAP actually is. The [supermarkets] need safe food and they need accountability and that. Well, that is fine; but what I find interesting is the supermarkets using that as leverage. Are they saying to us, as an organisation: "You meet those standards [and] we will take your fruit. We will sell it for a premium and da, da, da"? And what I would like to know, on the other hand, are they turning their back a little bit and saying "Well, hey, we will source some cheaper product from that country and we will still buy it and we will still sell it but the standards of securing it might not be as good." (Green)

---

5 Hi-Cane® is a toxic 'bud-burst' agent currently permitted under Integrated (it does not persist in the harvested fruit), but strictly prohibited under organic management systems. It is widely rumoured that EurepGAP will eventually prohibit Hi-Cane®, resulting in substantial yield losses.

Some growers also question the fairness of a system that could be exploited by lower cost producers or by the importing states themselves.

> Female: If things like ... EurepGAP ... which has been to a large degree ... it is a marketing response in terms of what the markets require in terms of traceability ... and that's a pain. It's an annual thing but quite pedantic and that is a pain, but it is a cost of doing business. You know it is.

> Male: I wish our competitors had to do the same amount ... be controlled by the same regulations and standards, the [in]equality that we have, that ... would make a difference. (Gold)

As a whole, these concerns encourage many orchardists to question whether the promised benefits of EurepGAP compliance compensate for the real and potential costs of submitting to outside control. This is especially true because the orchardists do not recognise any direct monetary benefits to their participation in the programme.

In summary, however, all these comments (and other interview responses they represent) indicate a strong level of agreement among orchardists that the new audit system has situated them within complex political relations with retailers, EU authorities and their competitor exporting countries. In general, the new situation is characterised by a loss of power and autonomy over on-orchard practices and questions about the appropriateness and fairness of audit requirements that are designed from Europe.

Discussion: Audit, Sustainability and Agri-Food Governance

Our assessment of grower-level response to the EurepGAP audit addresses several issues identified in the early theoretical literature on audit systems and sustainability. First, we can make some preliminary contributions to the broader discussion of the auditing of 'sustainability' through vehicles such as audit systems. Furthermore, it is evident that EurepGAP, while essentially similar to existing standards of practice, is viewed with some suspicion due to its organisational origins and the detail of its auditing requirements.

Outcomes for Orchard-Level Activities

The interview data provides comparable material to Guthman's (2000) critique of the real agro-ecological outcomes of certified organic practice in California, focusing instead on the real effects of production practices associated with the EurepGAP audit. In general, it has had no significant impact on the already well-established use of integrated management systems by kiwifruit producers. The similarity between EurepGAP and such systems was, in fact, the reason for ZESPRI's enthusiastic association with the EurepGAP alliance. The imposition of the audit has, however, raised the profile of policing mechanisms in regard to management standards and, in all practicality, removed the potential for non-compliance. The second main effect was the shift in the balance between on-orchard and in-office tasks for kiwifruit producers due to the increased paperwork requirements of the audit. Very few growers liked this, preferring to derive their esteem as growers from their performance in the orchard itself. Thus, the new audit appears to have minimal effect on the environmental performance of kiwifruit orchards.[6] It has, however, influenced compliance with local labour regulations, although the extent of positive impacts in the highly contentious territory of horticultural labour is disputed. Finally, audit compliance has altered the social dynamics through which orchardists define themselves, enjoy their work, or balance work and lifestyle choices. The similarities between EurepGAP and existing integrated management systems on kiwifruit orchards open speculation regarding the response of growers in other regions and sectors where this is not the case.

Outcomes for Agri-Food Politics and Relations

A second issue involves the level of grower autonomy relative to production practices. The interviewed growers indicate that EurepGAP has decreased such autonomy allowing them limited opportunity to 'bend' the standards to suit local management conditions. Again, the impact of this loss of autonomy is muted in an industry where management practices were already heavily proscribed by ZESPRI and where, significantly, growers' returns are currently buoyant.

A more complex range of dynamics emerges from this data in relation to growers' perceptions of alterations in some key power relations in agri-food systems as a result of EurepGAP auditing. Clearly EurepGAP, as a recognisably European product, is understood to be an expression of rising retailer power involving increased dictation of permissible practices. Growers also perceive EurepGAP as a potential tool with which retailers can play suppliers from different regions off one another – although, admittedly, this is hardly a dynamic unique to audit systems. Utilising its privileged position as the state-sanctioned monopoly exporter of New Zealand's crop, ZESPRI has placed EurepGAP (along with branding strategies, other quality measures, and new kiwifruit varieties) at the centre of an apparently successful strategy to secure their status as the elite supplier of kiwifruit to the European market (see Campbell *et al.* 2006a; Campbell *et al.* 2006b). The growers, however, have not uniformly incorporated this strategy and continue to see ZESPRI as an organisation that should advocate for growers, rather than as a market-strategy broker. Independent analysts of the industry generally agree that the ZESPRI strategy (including environmental auditing and alliance with EurepGAP) has been a resounding success and resulted in significantly higher returns for growers. This perception is not shared by all growers, and the increasing burden of paperwork and the perceived loss of managerial autonomy by growers appear to encourage increasing alienation from the industry's main organisation over the issue of audits.

---

6 As the ARGOS programme generates more data on the actual environmental performance of kiwifruit orchards under different management regimes, it will be possible to reflect more deeply on the degree to which sustainability claims and on-orchard outcomes are aligned with each other.

In conclusion, the clearest insight that can be drawn from the interviews is that EurepGAP not only reconfigures global-scale relationships between retailers and large corporate suppliers like ZESPRI (as argued in the published literature to date); it also initiates shifts and re-alignments between different institutional parties in the production regions themselves. Future research in the sector will focus on the extent to which this involves a re-alignment of grower loyalties towards grower representative organisations and the packhouses who act as intermediaries in the supply chain. If such a re-alignment exists, it likely has complex causes; but the arrival of environmental auditing and ZESPRI's commitment to EurepGAP has certainly played an important role.

EurepGAP clearly influences the style of work and management practice undertaken by orchardists – with a major shift towards more 'office work' and less 'orchard work'. As such, the audit exerts significant impact on how the orchardists view themselves as growers and their satisfaction in their work. However, there is no direct evidence that EurepGAP has significantly shifted sustainable practices, apart from a sense that compliance with the industry's prescription for 'best environmental practice' in kiwifruit production is increased. Yet, even in an industry where ZESPRI heavily prescribed integrated management standards in the mid-1990s and then translated such prescription to EurepGAP in 2003, the new system does indicate potential pressure points for warranting future investigation in other contexts. The successive systems of standards have circumscribed the range of grower management options in response to local production requirements. Such a bounding of acceptable practice is significant given the global debate regarding the importance of local, indigenous and context-specific knowledge for achieving sustainable outcomes in food production. Similarly, this new set of arrangements does clearly impose a European set of production standards into an Antipodean context. While ZESPRI, grower organisations and packhouses have all contributed to modifying EurepGAP requirements to suit New Zealand conditions, this has been achieved solely by means of ZESPRI's highly privileged position as the organisation leading the design of EurepGAP's standards for kiwifruit. As Campbell (2005) suggests, if there are challenges evident for making EurepGAP sustainable for kiwifruit production in New Zealand, these may be magnified many times in other export zones lacking the power, positioning or influence of ZESPRI in the world kiwifruit industry.

References

Buck, D., Getz, C. and Guthman, J. (1997), 'From farm to table: the organic vegetable commodity chain of Northern California', *Sociologia Ruralis*, vol. 37, pp. 3–20.

Busch, L. and Bain, C. (2004), 'New! Improved? The Transformation of the global agrifood system', *Rural Sociology*, vol. 69, pp. 321–46.

Campbell, H. (2005), 'The rise and rise of EurepGAP: The European (re)invention of colonial food relations?', *International Journal of Sociology of Agriculture and Food* 13, <http://www.csafe.org.nz/ijsaf/archive/vol13(2)_05/ campbell_dec05. pdf>.

Campbell, H. and Coombes, B. (1999), "Green protectionism' and organic food exporting from New Zealand: crisis experiments in the breakdown of Fordist trade and agricultural policies', *Rural Sociology*, vol. 64, pp. 302–319.

Campbell, H. and Le Heron, R. (2007), 'Supermarkets, producers and audit technologies: the constitutive micro-politics of food, legitimacy and governance' in Lawrence, G. and Burch, D. (eds), *Supermarkets and Agri-Food Supply Chains: Transformations in the Production and Consumption of Foods*, Edward Elgar, London, pp. 131–153.

Campbell, H. and Liepins, R. (2001), 'Naming organics: understanding organic standards in New Zealand as a discursive field', *Sociologia Ruralis*, vol. 41, pp. 21–39.

Campbell, H., Fairweather, J. and Steven, D. (1997), Recent Developments in Organic Food Production in New Zealand: Part 2, Kiwifruit in the Bay of Plenty', Studies in Rural Sustainability No. 2, Department of Anthropology, University of Otago, Dunedin, New Zealand.

Campbell, H., Lawrence, G. and Smith, K. (2006a), 'Audit cultures and the Antipodes: the implications of EurepGAP for New Zealand and Australian agri-food Industries', in Murdoch, J. and Marsden, T. (eds), *Between the Local and the Global: Confronting Complexity in the Contemporary Agri-Food Sector*, Research in Rural Sociology and Development 12, Elsevier, Oxford, pp. 69–95.

Campbell, H., McLeod, C. and Rosin, C. (2006b), 'Auditing sustainability: the impact of EurepGAP on organic exporting from New Zealand', in Holt, G. and Reed, M. (eds), *Organic Agriculture: A Sociological Perspective*, CABI, Oxon, pp. 157–73.

EurepGAP (2001), '*EurepGAP Protocol: Fresh Fruit and Vegetables*', Version, September 2001, FoodPLUS: Cologne, <http://www.eurepgap.org/fruit/ Languages/English/documents.html> (documents page, accessed 5 October 2001).

EurepGAP (2004a) '*General Regulations: Fruit and Vegetables* ', Version 2, January 2004, FoodPLUS: Cologne, <http://www.eurepgap.org/fruit/Languages/English/ documents.html> (documents page, accessed 9 March 2005).

EurepGAP (2004b), '*EurepGAP Control Points and Compliance Criteria*', Version, January 2004, FoodPLUS: Cologne, <http://www.eurepgap.org/fruit/Languages/ English/ documents.html> (documents page, accessed 9 March 2005).

EurepGAP Newsletter (2003–04), FoodPlus: Cologne <http://www.eurepgap.org/ fruit/Languages/English/publications.html>.

Guthman, J. (1998), 'Regulating meaning, appropriating nature: the codification of California organic agriculture', *Antipode*, vol. 30, no. 2, pp. 135–54.

Guthman, J. (2000), 'Raising organic: an agro-ecological assessment of grower practices in California', *Agriculture and Human Values*, vol. 17, pp. 257–66.

Guthman, J. (2004), *Agrarian Dreams: The Paradox of Organic Farming in California*, University of California Press, Berkeley.

Hall, A. and Mogyorody, V. (2001), 'Organic farmers in Ontario: an examination of the conventionalization argument', *Sociologia Ruralis*, vol. 41, no. 4, pp. 399–422.

Howley, V. (1997), 'The peak of practices', *Fresh*, September 1997.

Larner, W. and Le Heron, R. (2002), 'From economic globalisation to globalising economic processes: towards post-structural political economies', *Geoforum*, vol. 33, no. 4, pp. 415–19.

Le Heron, R. (2003) 'Creating food futures: reflections on good governance issues in New Zealand's agri-food sector', *Journal of Rural Studies*, vol. 19, pp. 111–25.

Morris, C. and Winter, M. (1999), 'Integrated farming systems: the "third way" for European agriculture?', *Land Use Policy*, vol. 16, pp. 193–205.

Shore, C. and Wright, S. (1999), 'Audit culture and anthropology: neoliberalism in British higher education', *Journal of the Royal Anthropological Institute*, vol. 5, no. 4, pp. 557–75.

ZESPRI (2003), *ZESPRI Grower Requirements: 2003/2004 Season Crop*, ZESPRI International Ltd, Tauranga, New Zealand.

Chapter 7

Maintaining the 'Clean Green' Image: Governance of On-Farm Environmental Practices in the New Zealand Dairy Industry

Paula Blackett and Richard Le Heron

Introduction

This chapter discusses the influences which facilitated a shift in the governance of on-farm environmental management practices on New Zealand dairy farms from local government towards industry led self-regulation. It is argued that several key elements proved instrumental in driving governance change. They were: first, scientific evidence which supported claims that water quality in dairying catchments was poor; second, an increased public concern over poor water and habitat quality in rural waterways and environmental externalities of food production, and third, and most importantly, an industry desire to maintain the 'clean green' image of New Zealand dairy production in the face of scientific and public concern. In order to explore these factors, the discussion is focused around the 'dirty dairying debate' which in essence was a public debate, through the media, instigated by Fish and Game New Zealand (a nation-wide fishing and hunting recreation group) over on-farm environmental management of water and habitat quality issues. This debate exposed a complex bundle of issues involving scientific evidence, economic issues and social knowledge in both local and global context. At its heart, it centres around the collision of values and interests between various actors including local authorities (Regional Councils), hunting and fishing organisations (Fish and Game New Zealand), individual farmers, farmer lobby groups (Federated Farmers) and key players in the dairy industry. By examining the rational and supporting evidence behind Fish and Game New Zealand's arguments, historical water quality governance, the interplay of industry and regulatory actors and the role of the consumer it will become clear how the transition in governance began.

It is argued that industry regulation of dairy farming environmental practices is the most sensible and practicable alternative to the current situation. This is primarily because it sets up a framework to deal with the issues raised during the 'dirty dairying debate'. Moreover, it will better facilitate improved water and habitat quality while affording dairy companies with the evidence they believe necessary to protect the 'clean green' image of New Zealand dairy production.

Origin of the 'Dirty Dairying Debate'

In June 2001, Fish and Game New Zealand placed classified advertisements in many local papers proclaiming that 'dirty dairying was a fact'. Through this advertisement, they insinuated that dairy farmers and the industry in general were degrading national water resources to make a profit. Furthermore, they proposed that resource management authorities (Regional Councils) were ineffectively mitigating the impacts of on-farm practices on water and habitat quality (Wybourne 2002). These claims were fuelled by concerns about waterway condition at a time of continual intensification on existing properties and expansion into new areas through the conversion of traditionally sheep and beef farms. Not surprising, given that between 1990 and 2003 the number of cows milked rose from 2.5 to over 3.5 million. Evidence used to support the argument was persuasive. It included a video of cows crossing a river and the associated downstream effects (Court 2002) and Local Authority reports on poor compliance with existing environmental requirements (MacBrayne 2002), and in particular, a review on the effects of agriculture on waterways prepared by a reputable scientific research organisation.

Scientific evidence generally supports the arguments presented by Fish and Game that dairy farming practices are affecting water and habitat quality. The two major reviews on the effects of agriculture on freshwater systems, produced in the last decade, report that small streams in dairy farming areas are in very poor condition (Smith *et al.* 1993, Parkyn *et al.* 2002). Poor condition is defined as high nitrogen and phosphorus concentrations, low dissolved oxygen levels and high counts of faecal coliform (mammalian intestinal bacteria) (Parkyn *et al.* 2002). When a waterway has these characteristics there are impacts on biodiversity, aesthetics and human use of waterways for recreation (fishing, swimming, kayaking for example) and as a source of drinking water.

On-farm practices that have been implicated as potentially influencing water and habitat quality include clearing riparian vegetation and stream channelisation (Quinn and Wilcock 2002), stock access to waterways (Reed 2003), stock crossing waterways (Davies-Colley *et al.* 2002), run-off from farm races and stand-off and feed pads, and dairy shed effluent discharges (Hickey *et al.* 1989).

In 2001, only dairy shed effluent (point pollution source) discharges were regulated while all of the other activities (diffuse pollution sources) were managed voluntarily by individual farmers. A key objective of the debate appeared to be lobbying local authorities to further regulate all on-farm environmental practices. Regional Councils were Fish and Game's primary targets because they are charged with environmental management under the New Zealand Resource Management Act 1991. Fish and Game believed that further regulation of dairy shed effluent and diffuse inputs by Regional Councils would more adequately protect water and habitat quality.

By forcing other organisations to contest the issue in the media, rural water and habitat quality was placed in the public and political arena. A number of other actors were drawn into the debate including farmers, farmer lobby groups and dairy companies.

Farmers felt unfairly blamed by the accusations as many thought by complying with Regional Council requirements they were not impacting the environment. In addition, there was some attempts by farmer groups to widen the debate to the effects of urban populations or other industries on waterways (Blackett 2004).

To the dairy companies, the debate represented a threat to the 'clean green' image of New Zealand food production, a key product differentiation marketing tool – Fish and Game were accused of committing 'economic treason by putting the export industry at risk' (Lee 2001). There are several dairy companies in New Zealand, of which Fonterra is by far the largest, collecting milk from 95 percent of dairy farmers nationwide. Fonterra is the world's largest dairy exporter, one of the largest dairy companies by value and by volume of milk processed. Overall, the New Zealand dairy industry is worth $NZ12.3 billion per year, predominantly from export earnings (Fonterra.com). It is clear why the 'dirty dairying debate' was taken so seriously.

An Industry Perspective

The dairy industry has two key concerns, first, that consumers will avoid products because they perceive production is unsustainable, and second that environmentally based tariffs may be imposed on exports and imports.

Production of food has come under increasing scrutiny from a public concerned over food safety and social and environmental externalities of food production systems. A Ministry for the Environment (2001) study presented information on the environmental sensitivity of New Zealand's overseas customers and the importance of the 'clean green' image. This claimed that consumers from lucrative North American, European and United Kingdom markets could select against New Zealand products should the 'clean green' image be framed as closer to 'filthy and brown' as suggested by Fish and Game (Fish and Game 2001). Although the 'dirty dairying debate' presented a key threat to the 'clean green' perception it is not the only recent challenge. For example, a British documentary raised questions over the 'clean green' claims of the New Zealand industry, and made particular reference to sensitive environmental and animal welfare issues (Dairy Exporter 2002). As long as examples of 'poor' management exist the threat remains. Effect on the milk commodity market in other parts of the world is less clear.

A further industry concern is a possible implementation of dairy exports tariffs based around environmental management standards (Blackett 2004). Inspectors from the European Union already randomly visit farms to check on a host of export standards (Mountfort 2002), it would not be difficult to extend the scope into on-farm environmental management practices. This could be very costly if the industry had no means to defend its environmental record.

The value of the 'clean green' image emerges clearly if it is viewed at the individual farmer level, where an average dairy farmer could loose an estimated $18,000 to $49,000 if the image were tarnished (Pickering 2002). This does not take into account value derived from any other industry, for example tourism, or sheep and beef markets. Overall the risks to the dairy industry associated with not acting over on-farm environmental practices are considerable, particularly given the scientific evidence supporting Fish and Game's argument.

Governance of Dairy Farming Practices

From a historical perspective, statutory regulation is the first type of method put in place to control an activity (Dryzek 1997), however, there are a number of other options available including economic instruments, voluntary actions and industry regulation. To date, a mix of these approaches has been used to manage on-farm environmental practices.

Historical Regulation of On-Farm Environmental Practices

Regulation of point source discharges in New Zealand began with the enactment of the Water Pollution Act 1953. Although this act had no impact on dairy shed effluent disposal it marked the beginning of a way of thinking about regulation as a means to control discharges to water. In the 1970s regulation of dairy shed effluent began, influenced by the Water and Soil Conservation Act 1967, raising public environmental awareness (Dryzek 1997) and increases in national dairy herd size (Blackett 2004). Controls were initiated and enforced at the regional level and were based on implementation of a specific technology – the two stage oxidation pond. Discharges were allowed provided a farmer met the existing design standard (Ministry of Works 1972). The effects of the treatment system on the receiving environment were not assessed (Blackett 2004).

Research, by Hickey *et al.* (1989) illustrated that the standard design was not as effective at treating the waste stream as expected and were resulting in likely impacts in freshwater environments. This research, coupled with evidence that dairy farming was expanding into new regions and intensifying and the enactment of new environmental legislation (Resource Management Act 1991), led to three significant shifts in effluent regulations. First, standard pond design was revisited, with size of the aerobic pond increased by 40–46 percent in an attempt to improve functioning (Sukias *et al.* 2002). Second, alternative treatment designs or pond improvements were investigated, for example the advanced pond system (Craggs *et al.* 2000), wetland add-ons (Tanner and Kloosterman 1997) or additional aeration (Sukias *et al.* 2002). This is still a very active research area. Third, several regional authorities shifted towards a preference for land application of waste in an attempt to recycle nutrients to the land and avoid the problem of discharges to water.

Under the current environmental legislation (Resource Management Act 1991) Regional Councils are able to specify how dairy shed effluent is to be managed in their particular region. They are given reasonable freedom to structure requirements as they consider appropriate, including using an environmental effect based approach. However, there is still a strong reliance on regulations which still essentially centre around use of particular treatment technologies in spite of little information on the actual effects of different treatment technologies on receiving waters (Blackett 2004).

It is interesting to note that in spite of the potential impacts on receiving waters the type of treatment system was not raised in the 'dirty dairying debate'. The focus was simply on monitoring and enforcement issues.

Monitoring and Enforcement Practices

To be effective any rules must be enforced (McKean 1992) and the RMA (1991) provides several enforcement alternatives including, abatement notices (warning), infringement notices (fines) and prosecution. One of the key elements of the 'dirty dairying debate' was Fish and Game's belief that Regional Council monitoring and enforcement process were ineffective and not adequately protecting waterways. Farmer non-compliance with regional rules was frequently quoted as between 5 percent and 80 percent depending on the region in question (MacBrayne 2002). However, this should be treated carefully because it is not clear how information was collected and what the reasons for the non-compliance were. In addition, fines and prosecutions were typically low in spite of high non-compliance rates.

Prior to 2004, Regional Councils tended to rely on either spot checks or systematic farm by farm checks of system compliance by on-the-ground personal. Checks may be carried out by Council employees or contractors (Blackett 2004). However, the 'dirty dairying debate' has had an impact on monitoring and enforcement practices. In the last two years Environment Waikato (Waikato Regional Council) have moved from contracting out the effluent systems checks to employing more in-house field inspectors. For the 2005/06 year, they introduced a helicopter fly over to more cost effectively assess compliance with the permitted activity of effluent irrigation. Any problems spotted during the fly over are followed up by field inspectors and may result in an enforcement action if deemed appropriate. Out of 600 farms investigated in 2005/06 57 received $750 instant fines while 83 were given abatement notices, translating to around 56 percent compliance (Franks 2006). After each catchment fly over, Environment Waikato held local meetings to discuss issues and answer questions. Overall the monitoring approach seems to be working as each progressive catchment has more compliant farmers as word of the helicopter inspection spreads (Franks 2006).

Voluntary Regulation of Diffuse Inputs

In contrast to point source discharges of dairy shed effluent, diffuse inputs have never been subject to regulation, instead Regional Councils tend to rely on voluntary action to mitigate the effects of dairy farming practices. Adoption of on-farm environmental management is generally encouraged through providing information, using financial incentives or promoting collective community based action.

The information deficit model forms the underlying concept behind provision of information to farmers as a means to effect behaviour change. It assumes a lack of appropriate information is the barrier to rational decisions about on-farm environmental practices, and proposes that correcting the deficit alters behaviour (Owens 2000). There is little doubt that information is an important part of making a decision. However, Vanclay and Lawrence (1994) suggest for many farmers not adopting an environmental practice may also be a rational option. In fact, the model fails to take account of the economic, social, political and cultural contexts and constraints of everyday life through which new information is evaluated and assimilated (Owens 2000). Regional Councils tend to subscribe to this model and often create brochures, booklets and information packages promoting the changes they desire incorporating targeted details on the benefits and costs of certain activities. It is probably more effective where regulation applies (i.e., dairy shed effluent disposal) than for diffuse pollution management on dairy farms (Blackett 2004).

Financial incentives in the form of subsidies, are commonly used to encourage retirement and planting of stream margins. Most Regional Councils run a cost share scheme where planting and/or fencing costs in priority catchments are partially funded. Schemes have been quite successful in some areas, but the farmer still carries some of the costs.

Landcare or streamcare groups form the core of current voluntary collective action to mitigate the effects of agricultural practices on waterways in both New Zealand and Australia (Byron and Curtis 2001). Action through self regulating groups is considered to be a more democratic, participatory and inclusive means of approaching the problem which hinges around the negotiation of shared environmental values and goals (Owens 2000). Groups may be completely self organising or have a Regional Council/New Zealand Landcare Trust facilitator who acts as a motivator and information supplier but gives participants control over their own learning and actions (Wilson 1997). Landcare or Streamcare groups have had some successes, which are widely promoted by Regional Councils and the New Zealand Landcare Trust. However, Ritchie (1998) believes they have not proved an effective way to deal with diffuse pollution in New Zealand because they are frequently under-resourced and under-financed. Moreover, many farmers do not participate in such groups (Blackett 2004).

The success of voluntary mitigation of diffuse pollution from dairy farming practices was called into question as part of the 'dirty dairying debate'. Fish and Game believed that all mitigation should be regulated than voluntary because it would ensure protection of waters rather than leaving choices to individual farmers. Proponents of voluntary adoption would argue that change would happen – eventually. However, whether or not this would be in time to defend the 'clean green' image given the existing scientific evidence is questionable.

In Summary

Overall, management of on-farm environmental practices in New Zealand through regulation of dairy shed effluent or voluntary mitigation of diffuse pollution sources were not adequately protecting water and habitat quality. This is supported by scientific evidence in that water and habitat quality of dairy farming catchments was poor. In other words the 'dirty dairying' debate's challenge to the 'clean green' image was not unjustified. It became a matter, at least in the eyes of major organisations implicated and interlinked in dairy effluent management, of how to move forward, in particular who should be responsible for governing the industry and how should it be done.

Interactions between Dairy Industry Actors

In the past, different dairy industry actors (i.e., Fonterra, Dexcel, Regional Councils, agricultural consultants) have had quite different goals and did not often associate with each other. As a consequence, environmental management initiatives were uncoordinated, fragmented and lacked any cohesion across industry actors. Each agency tended to peruse their own goals to the general detriment of farmers' ability to make on-farm environmental management decisions.

In the Waikato Region, Environment Waikato set the regional rules for dairy shed effluent disposal, monitored compliance and produced information detailing the environmental standards farmers should meet. Support for voluntary action was through various publications, financial incentives for riparian management and facilitation of Streamcare and Landcare groups. They saw themselves as setting and monitoring the rules but not as an agency to help farmers with on-farm decisions because they felt it is important to allow farmers flexibility in how to comply with regional rules. Other extension agents and environmental consultants were considered responsible for provision of technical support and advice. In general, Regional Councils do appear to struggle with the tensions between establishing and enforcing rules and providing assistance to farmers on how to comply.

By contracting out the regular effluent inspection system Environment Waikato had limited contact with farmers. Farmers perceived them as the regulators but were afraid to ask for advice because of potential repercussions of farm visits – unfortunate because who better to talk through options with.

Dexcel (dairy extension and research organisation) were viewed by farmers as production specialists and not considered an authority on environmental advice (Blackett 2004). However, Dexcel do have links with farmers through their consulting officers via the farm discussion group system. Similar to Dexcel, private agricultural consultants deal mainly with production issues and have extensive client networks but limited contact with any other agency (Blackett 2004).

The result of low levels of contact and information exchange between these organisations did not help promote adoption of mitigation technologies on individual farms. In fact, it probably only added more complexity to issues around waterway management at the farm level.

With respect to future regulation of dairy farming practices, Regional Councils expressed a preference for the status quo. They held to the belief that the voluntary practices in place would produce results in the long term. In addition, there was concern over how changes would be resourced to ensure effective monitoring and enforcement. Farmer lobby groups supported this position and wished to preserve as much individual choice as possible. It was clear, that changes significant enough to defend against the 'dirty dairying debate' were unlikely to arise from these quarters. At the end of the day Fonterra chose to self-regulate before the government intervened to defend the 'clean green' image themselves.

Towards Industry Regulation – Evolution of the Clean Streams Accord

The reasons for a shift in governance towards an industry regulated approach to managing the environmental externalities of dairy production are due to several key influences. First, that dairy farming practices were impacting on water and habitat quality. Second, that adequate protection of water resources was not being achieved through regulation of point source discharges and voluntary mitigation of diffuse pollution. Third, that the previous two statements could affect sales of dairy products to overseas markets.

Industry control of on-farm environmental management practices evolved rapidly with around two years between the instigation of the 'dirty dairying debate' and the signing of the Clean Streams Accord (The Accord) in May 2003 (Fonterra Co-operative Group *et al.* 2003). Fonterra's Clean Streams Accord complements and sits parallel to Regional Council regulation of dairy shed effluent discharges, promotion of diffuse source management by financial incentives and voluntary action. Moreover, it was designed specifically to pre-empt issues raised by the 'dirty dairying debate' by committing to reduce the environmental impacts of dairy farming on water and habitat quality.

The Accord sets a series of timeframes for farmers to address fencing of waterways and significant wetlands, replacement of stock river crossings with bridges, adoption of nutrient management practices and compliance with dairy shed effluent resource consents (Table 7.1) (Fonterra Co-operative Group *et al.* 2003). Farmers who do not meet the requirements of The Accord will not have their milk accepted.

To support The Accord, Dairy Insight (the organisation response for distribution of research funds obtained from a farmer levy) formed the Dairy Environment Review Group to create a research investment strategy to achieve the dairy industry's environmental goals over the next 10 years (Dairy Environment Review Group 2006). This document identifies key priority areas and overall, it signals an industry intention to become increasingly involved with Regional Councils, researchers, and farmers to achieve improvements in management of on-farm environmental practices. It builds on The Accord, but does not indicate how any of the objectives will be achieved – presumably this is the next step.

Implementation of the Clean Streams Accord

The fulfilment of The Accord and closer relationships between the dairy industry, researchers, Regional Councils and farmers is potentially an ideal solution to governance issues over on-farm environmental practices. Moreover, it would provide evidence to satisfy any challenges to the 'clean green' image of dairy production in New Zealand by addressing water and habitat quality issues. However, to make a decision to reduce the impacts of dairy farming practices on freshwater environments and set broad objectives is only the first step towards improved outcomes. Continued alignment and co-ordination of dairy industry actors to better assist farmers achieve real change in on-farm environmental decision making is very important. Furthermore, that good science and practicable solutions must underlie mitigation technologies used to achieve the objectives of The Accord. Mitigation

Table 7.1   The Clean Streams Accord dairy industry goals and target timeframes

Industry goal	Target time frame
Cows fenced out of streams, rivers and lakes and their banks.	50% of streams, rivers and lakes fenced by 2007, 90% by 2012.
Farm races include bridges or culverts where stock regularly cross a watercourse.	50% of regular crossing points have bridges or culverts by 2007, 90% by 2012.
Farm dairy effluent is appropriately treated and discharged.	100% of farm dairy effluent discharges to comply with Regional Council rules immediately.
Nutrients are managed effectively to minimise losses to ground and surface waters.	100% dairy farms to have adopted nutrient budgeting systems by 2007.
Existing regionally significant or important wetlands are fenced and natural water regimes are protected.	50% of regionally significant wetlands to be fenced by 2005, 90% by 2007.

Source: Derived from Fonterra Co-operative Group *et al.* 2003.

strategies may need regular reviews to ensure that the management strategy is actually protecting receiving waters. In particular, dairy shed effluent discharge rules may need revisiting.

A simple idea of protecting the 'clean green' image of New Zealand dairy production has required, and will continue to create, some very complex changes in the way Fonterra, Regional Councils, Dexcel, scientists, and agricultural consultants interact with each other and farmers.

Adoption of The Accord and the Dairy Industry Environmental Management Strategy has improved relationships at the more strategic level with inter-agency co-operation and dialogue. Nevertheless, there are still a number of issues around conflicting inter and intra-agency goals and responsibilities, particularly around production goals and farmers' assistance. Resolution of these issues and a robust monitoring and enforcement programme will be fundamental to The Accord's success.

Industry Regulation as a 'Sensible' Alternative

By choosing to create 'The Clean Streams Accord' Fonterra has effectively recognised the impacts of farming practices on waterways, assessed the limitations of current governance structures and formulated new policies to protect their interests. A desire to improve environmental performance, thus public perceptions, of the industry or counter concerns of future government regulation are common reasons for industry self-regulation initiatives (Labatt and Maclaren 1998). Each is clearly present in the factors driving The Accord.

Clear advantages of industry self-regulation are that administration and enforcement costs are not borne by the wider community and compliance rates are typically good because of the penalties of non-compliance. In this respect, Fonterra has considerably more power than local authorities because they can require all farmers to conform to The Accord as a condition of milk purchase. Farmers do not have many (if any) alternatives because of Fonterra's dominance in the New Zealand dairy processing industry.

In formulating The Accord, Fonterra worked closely with local government which has facilitated alignment of environmental targets. This coordinated approach is important because it attempts to limit duplication of environmental standards between different regulators. At present regulations are compatible and complementary with local authority and industry actors working together and managing separate areas of compliance. Local authorities still set the point source discharge rules and promote and support action on diffuse pollution control.

All of these features make The Accord a sensible and practical alternative to pure local authority governance. Moreover, it will potentially achieve improvements in water and habitat quality more rapidly than the previous governance structure because farmers are required to undertake certain environmental management strategies as a condition of milk supply.

Conclusions

This chapter has argued that a shift in the governance of on-farm environmental practices towards industry self regulation was driven by three key factors and is a sensible step towards protecting the 'clean green' image of New Zealand dairy production. It represents perhaps the best solution in the context because, prompted by the 'dirty dairying' campaign and ensuing debate, the industry has begun to address its own issues.

First, The Accord recognised the potential impact of dairy farming on waterways and established broad time bounded rules around on-farm management practices. Scientific evidence over the condition of New Zealand waterways played a significant role in both the instigation of the debate and establishing conditions of The Accord. Furthermore, science highlighted the inadequacy of current point source discharge regulation and voluntary mitigation of diffuse pollution which moved thinking towards another mix of governance strategies. An industry initiative was necessary because Regional Councils failed and were unwilling to resource further regulation themselves, but were willing to support industry regulation.

Secondly, The Accord acts to counter heightened public concern over the environmental impacts of dairy farming by providing a clear set of targets to reduce impacts on waterways. In addition, it has brought many industry actors closer to common ground and resulted in some steps towards resolving past differences and building relationships. More importantly, it has provided the building blocks for a common set of initiatives between various actors. This has reduced arguments in the media and could potentially increase public confidence that the issues are being addressed.

Thirdly, The Accord is evidence that dairy production in New Zealand is sensitive to environmental concerns. Moreover, it illustrates how the industry is actively seeking to reduce impacts on water and habitat quality. This may prove invaluable in future markets, particularly if monitoring and enforcement is effective and documented. However, how this Accord proceeds and its success over mitigating the impacts of dairy farming on waterways will be watched with interest. Failure to achieve success may push the industry into another 'dirty dairying debate'; perhaps with a different mix of antagonists and more international scrutiny of the New Zealand dairy environmental programme.

In the present conditions and given the differing abilities of organisations to incentivise and change behaviour though for positive and negative pressures, the Clean Streams Accord is the mostly likely governance structure to improve water and habitat quality given the current institutional arrangements in New Zealand. This is because the dairy industry has a strong and ongoing interest in maintaining the 'clean green' image of New Zealand food production.

Acknowledgements

This research was funded by the New Zealand Foundation for Research Science and Technology Contract C01X0022 and administered through the Institute of Atmospherics and Aquatic Science (a joint research collaborative between The University of Auckland and NIWA). Paula expresses her appreciation to Neels Botha, Toni White, John Quinn and Lyn Hanes and two anonymous referees for providing useful comments on this chapter. Richard expresses his appreciation for extensive dialogue with John Quinn and others at NIWA.

References

Blackett, P.E. (2004), 'Biophysical and institutional challenges to management of dairy shed effluent and stream management practices on New Zealand dairy farms', Unpublished PhD thesis, The University of Auckland, New Zealand.

Byron, I. and Curtis, A. (2001), 'Landcare in Australia: burned out and browned off'. *Local Environment*, vol. 6, pp. 311–26.

Court, J. (2002), Investigation of dairy farming practices in the Buller catchment, Ministry for the Environment, Wellington.

Craggs, R.J., Tanner, C.C., Sukias, J.P.S. and Davies-Colley, R.J. (2000), 'Nitrification potential of attached biofilms in dairy farm waste stabilisation ponds', *Water Science and Technology*, vol. 42, pp. 195–202.

Dairy Environment Review Group (2006), 'Dairy industry strategy for sustainable environmental management', *Dairy Insight*, www.dairynz.co.nz.

Dairy Exporter (2002), 'Insight into images on television abroad', *Dairy Exporter*, September 2002, p. 14.

Davies-Colley, R., Nagels, J.W., Smith, R., Young, R. and Phillips, C. (eds) (2002), 'Water quality impact of cows crossing an agricultural stream, the Sherry River,

New Zealand', in *Proceedings of the Sixth International Conference on Diffuse Pollution*, Amsterdam, pp. 671–8.

Dryzek, J.S. (1997), *The Politics of the Earth: Environmental Discourses*, Oxford University Press New York.

Fish and Game (2001), 'Dirty dairying' condemned by Fish and Game New Zealand, Press Release, 27 March.

Fonterra Co-operative Group, Regional Councils, Ministry for the Environment and Ministry of Agriculture and Forestry (2003), Dairying and Clean Streams Accord, <http://www.mfe.govt.nz/issues/land/rural/dairying-accord-may03.pdf>.

Franks, B. (2006), Dairy monitoring of permitted activity farms 2005/06, Environment Waikato, Hamilton New Zealand.

Hickey, C.W., Quinn, J.M. and Davies-Colley, R.J. (1989), 'Effluent characteristics of dairy shed oxidation ponds and their potential impacts on rivers', *New Zealand Journal of Marine and Freshwater Research*, vol. 23, pp. 569–84.

Labatt, S. and Maclaren, V.W. (1998), 'Voluntary corporate environmental initiatives: a typology and preliminary investigation', *Environment and Planning C: Government and Policy*, vol. 16, pp. 191–209.

Lee, A. (2001), 'Fishing, hunting lobby group sees dairying as fair game', *Dairy Exporter*, September 2001, p. 136.

MacBrayne, R. (2002), 'Farming effluent discharge brings $6750 fine', *The New Zealand Herald*, 22 August, Auckland.

McKean, M.A. (1992), 'Success on the commons: A comparative examination of institutions for common property resource management', *Journal of Theoretical Politics*, vol. 4, pp. 247–81.

Ministry for the Environment (2001), *Valuing New Zealand's clean green image*, Ministry for the Environment, Wellington.

Ministry of Works (1972), *Lagoon treatment of farm wastes*, Public Health Engineering Division, Civil Division Ministry of Works, Wellington.

Mountfort, M. (2002), 'Environmental, welfare assessments alongside farm dairy inspectors', *Dairy Exporter*, September 2002, p. 16.

Owens, S. (2000), 'Engaging the public: information and deliberation in environmental policy', *Environment and Planning A*, vol. 32, pp. 1141–8.

Parkyn, S., Matheson, F., Cooke, J. and Quinn, J. (2002), Review of the Environmental Effects of Agriculture on Freshwaters, NIWA, Hamilton.

Pickering, S. (2002), 'Farmers cleaning up their own backyard', *Waikato Times*, Hamilton.

Quinn, J. and Wilcock, B. (2002), 'Clean dairying, clean waterways; true sustainability', *AgScience*, September 2002, pp. 4–5.

Reed, T. (2003), 'Macroinvertebrate assemblage change in a small Eastern Oregon stream following disturbance by cattle grazing', *Journal of Freshwater Biology*, vol. 18, pp. 315–19.

Ritchie, H. (1998), 'Beyond the fences: co-ordinating individual action in rural resource management through Landcare: a case study of managing non-point source discharges to water in Waikato, New Zealand'. Unpublished PhD Thesis, University of Western Sydney, Sydney.

Smith, C.M., Wilcock, R.J., Vant, W.N. and Cooper, A.B. (1993), Towards Sustainable Agriculture: Freshwater Quality in New Zealand and the Influence of Agriculture, Ministry of Agriculture and Fisheries, Wellington, New Zealand.

Sukias, J.P.S., Tanner, C.C. and Nagels, J.W. (2002), 'Outflow volumes and mass flow impacts on receiving waters from dairy farm pond systems', in New Zealand Water and Waste Association (ed.), *Fifth International Specialist Group Conference on Waste Stabilisation Ponds: Pond Technology for the New Millennium*, New Zealand Water and Waste Association, Auckland, New Zealand, pp. 801–8.

Tanner, C.C. and Kloosterman, V.C. (1997), Guidelines for Constructed Wetland Treatment of Farm Dairy Wastewaters in New Zealand, NIWA, Hamilton New Zealand.

Vanclay, F. and Lawrence, G. (1994), 'Farmer rationality and the adoption of environmentally sound practices; A critique of the assumptions of traditional agricultural extension', *European Journal of Agricultural Education and Extension*, vol. 1, pp. 59–90.

Wilson, P.A. (1997), 'Building social capital: A learning agenda for the twenty-first century', *Urban Studies*, vol. 34, pp. 745–60.

Wybourne, M. (2002), 'Dirty waterways', *The Daily Post*, 9 July, Rotorua.

Chapter 8

A Sustainable Fisheries Oasis? Strategy and Performance in the New Zealand Seafood Sector

Eugene Rees

Introduction

In 2002 Alastair MacFarlane (2002, 4) declared 'New Zealand will be among a very small number of developed country fisheries with stable and sustainable fisheries harvests – a sustainable fisheries oasis'. But what sort of oasis is it? Internationally, much analysis has been focused on the Quota Management System (QMS) itself; the 'nuts and bolts' of property rights reform and the scientific outcomes for fish stocks have been studied, yet changes to aqua-commodity production made possible by the QMS have passed largely unheralded. This chapter contributes to geographic enquiry by setting out these changes and moves enquiry beyond debates over equity that surround the implementation of the QMS and property rights in fisheries to explore how firms have gone about securing growth when operating in the New Zealand fishing sector. It provides a general account of their attempts of vertical integration and diversification into the processing part of the chain and beyond the sector. Interviews with 53 fisheries sector participants carried out in 2003–05, analysis of secondary data and a review of the relevant literature demonstrates that the QMS is far more complex in terms of its outcomes and operations than is recognised or accepted by fisheries economists and those espousing marketised rights-based reform of resource management and allocation.

Many people involved in fisheries are aware of the broad dilemma. Demand for fish has increased over the last three decades. Fisheries scientists tell us that fisheries crises result from poor stock assessments and a failure to appreciate the wider ecosystems that fisheries are situated in (Dayton *et al.* 2002). The solution, for them, rests in more and better science. A number of economic analysts argue that fisheries crises result from poorly constructed property rights and the markets in which these rights are traded. For them, the solution is simple: simultaneously modernising fishing, securing sustainability of harvesting wild resources and maximising economic performance rests on restructuring property regimes. Assuming profit maximising behaviour, analysts may make different conclusions because of different assumptions about population dynamics, the price of fish, species and so on, but the underlying ethos remains the same – open access to a fishery results in inefficiency and tragedy (Anderson 2002).

Notably, some anthropologists, sociologists and economists argue that fisheries economists' ideas concerning the applicability of property regimes are at best ill-conceived, their remedies so disastrous as to inevitably lead to new crises (Hutchings 1999; Ludwig *et al.* 1994; Wright 2004; *inter alia*). This literature maintains that many of these communities have long established, and often complex community rules of management and regulation. Conventional economic theory quite simply does not fit the empirical observations and adds little to our understanding of the structure, organisational forms and dynamics of communities (see for example, Newell and Omer 1999).

Value based tensions arise as analysts privilege a specific mode of regulation, yet all remain partial accounts. When economists act to facilitate 'proper' capitalist management and exploitation of the resource, they often omit or assume away the organisation and performance of the industry in their analysis. When anthropologists or sociologists recommend co-management and customary or traditional resource rationing practices they often omit globalising processes and the ways in which these are played out at regional levels from the discussion. Neither places the key actor – the fishing enterprise at the centre of the discussion, or acknowledges the political economy of firms. Neither can establish the extent to which fishing tragedies are a firm problem or a fisheries problem. Sustainable development has proved elusive.

Charting a New Course

An emerging literature explores individual firm behaviour and wider sector adjustment as management regimes change. It acknowledges the culture of fishing, the wider community in which fishers are situated and the economic imperatives and institutional culture of fishing companies to which previous literatures only allude. Arbo and Hersoug (1997) argue that local change is an outcome of emerging regimes of regulation across a number of scales, the development and diffusion of technology, the emergence of multinational corporations, the power of firms acting at the market end of commodity chains, the general liberalisation of trade and the movement of capital. Utilising field dynamics Fløysand and Lindkvist (2001) demonstrate that firms are simultaneously part of a community field, a region field and a corporation field. They contend that, despite globalisation, culturally embedded knowledge at the regional, community and enterprise levels explains observed development in fisheries. Mansfield's (2001a; 2001b) study of the United States North Pacific fisheries explores the role of political and economic actors in constructing a fishery that became simultaneously national and transnational as local enterprise constructed a series of joint ventures and strategic alliances within the fishery. More recently Mansfield has turned her attention to privatisation and neoliberal regulation and in doing so recognises the importance of context (place) when considering the outcomes of neoliberal regulation (Mansfield 2004a; 2004b). All ground their research in the local while acknowledging the wider political economy in which fishers operate. Such an approach enables an investigation into the performance and expectations of fisheries management for sustainable growth of New Zealand's seafood sector.

The New Zealand Context

In 1986 New Zealand established the Individual Transferable Quota (ITQ) system under the QMS regime in response to perceived crises of overcapacity and over-fishing in the inshore fisheries and fears of impending crisis in deep-water fisheries. Now quota systems are very common (and contentious). Prior to this the New Zealand fishing sector had experienced cycles of investment followed by stagnation and later disinvestment as resources were placed under pressure. Significantly, the State, through legislation and regulation controls, played an active role in shaping industry behaviour.[1]

New Zealand's QMS is a dynamic institution that has undergone many refinements but the basic structure – setting TACC[2] and leaving the market to determine the most profitable allocation of fishing effort – remains. The QMS is intended to meet four key objectives: stock sustainability; industry growth; New Zealandisation and increased industry competitiveness. It does this by realigning industry through market signals, leaving only the most efficient fishers to extract the maximum value from fisheries. The intended outcome is a competitive, market driven industry that endeavours to extract maximum value from resources while ensuring sustainability and resources for future generations.

The QMS does not dictate fleet structure or distribution of catch effort; nonetheless it does provide a context within which these decisions are made. The QMS controls fishing effort by limiting the amount of fish that can be caught while allowing quota holders to buy sell and lease their catching rights. To be successful quota has to be secure and durable over time, the rights garnered have to be incontestable and transferable so inefficient fishers may exit and efficient fishers can gain a larger share of the fishery. The QMS allows fishers to adjust their individual role within a fishery (Hersoug 2002). Quota encourages fishers to farm a resource rather than act as hunter/gatherers. Quota is expected to reduce the race to fish and commensurate bycatch associated with Olympic fisheries.

Realigning Production

A key expectation of the QMS was that there would be a realignment of production with seafood companies striving to maximise value through value-adding and more effective use of labour and capital. The QMS has resulted in a rationalisation of effort without explicitly regulating fleet capacity in the form of vessel numbers. There were fewer vessels operating in the fishery in 2002 than in 1984. Vessel numbers declined substantially in the inshore fishery where quota reductions were largest. Decline in deep-water vessels may be evidence of capacity creep and increased specialisation as more effective and efficient fishing techniques are employed. The mid-water fleet has remained relatively stable and is able to fill a number of roles, shifting fishing effort across a number of species and fishing areas. Equally, stability of the mid-water fleet is a function of vertical integration and consolidation as fewer firms fish

---

1 For a comprehensive history of New Zealand fisheries see Straker *et al.* (2002).
2 Total Allowable Commercial Catch.

longer with the same vessels. However, while absolute numbers have decreased the tonnage in some classes has increased and more than compensates for individual vessels removed in smaller classes (Connor 2001). However, quota buybacks and the QMS have led to rationalisation of vessels although arguably not of total catch capacity. Fishing has been tempered by a history of joint ventures and charter arrangements. Regardless, there has been a New Zealandisation of fishing effort. New Zealandisation is multi-layered and contested, but is taken as meaning building New Zealand capacity to the point where foreign capacity is only employed at the margins. While foreign vessel numbers have declined, charter vessels continue to play a role in high-volume deep-water fisheries and the seasonal squid fishery. These charter arrangements enable flexibility and efficiency gains as fishing companies are able to adjust capacity without directly impinging on their fleet.

The QMS and ITQs were expected to lead to increased employment in the seafood sector as New Zealand firms expanded into deepwater fishing, aquaculture and processing. There has been a reduction in numbers employed in fishing and an expansion in numbers employed in processing. Full-time equivalent employment has increased from 5,670 to 8,151 since the introduction of the QMS (Statistics New Zealand 2001). However, summary employment statistics in the fishing industry indicate contradictory trends. The removal of foreign vessels, expansion of deep-sea fishing and growth in aquaculture mean the fishing component remains at worst static and at best has undergone limited growth. Growth has been largely in processing and illustrates that the industry is value-adding to exports (Rees 2006). Employment has grown, but aggregate data obscures the detail of whether there has been a gendering of production, the nature of work done or when hours are worked.

There is an uneven geography of both investment and disinvestment as fishing and processing companies have rationalised effort. New Zealand's transition to the QMS, though arguably smooth at the national level, involved regional dislocation, with the collapse of fishing in Northland and the growth of fishing and processing activity in other regions. In 1981 the top four Territorial Land Authorities (Nelson City, Invercargill, the Far North District and Marlborough District) accounted for 26 percent of those employed in fishing (Table 8.1). Ten years later the Far North and Invercargill had been replaced by Tasman District and Christchurch, and they had grown to 36.9 percent of the labour force. In 2001 those regions were unchanged but had increased their share to 44.6 percent. The pre-eminent fishing region is Nelson, where share of employment has increased from 7.8 percent in 1981 to 15.7 percent of the total employed in 2001. Marlborough's growth is an outcome of significant expansion in the aquaculture industry and associated expansion of onshore processing facilities. Christchurch City has also increased its share of employment from 3.3 percent to 9.6 percent through the expansion of onshore processing. The 'top of the South' dominates employment in 2001.

Table 8.1   Employment concentration in the New Zealand fishing industry 1981–2001

	1981			1991			2001		
	Territorial Authority	% employed	LQ	Territorial Authority	% employed	LQ	Territorial Authority	% employed	LQ
C4	Nelson City	7.8	7.5	Nelson City	13.8	12.5	Nelson City	15.7	13.8
	Invercargill City	7.2	3.7	Marlborough District	9.1	8.6	Marlborough District	9.7	8.6
	Far North District	5.8	5.4	Tasman District	8.1	7.5	Tasman District	9.6	8.1
	Marlborough District	5.1	5.3	Christchurch City	6.0	0.7	Christchurch City	9.6	1.1
		26.0			**36.9**			**44.6**	
	Auckland City	4.8	0.5	Invercargill City	4.6	2.9	Timaru District	5.3	4.7
	Thames-Coromandel District	4.8	10.2	Thames-Coromandel District	4.3	7.9	Dunedin City	4.7	1.6
C10	Tasman District	4.3	4.4	Timaru District	4.0	3.4	Invercargill City	4.2	3.2
	Dunedin City	4.2	1.2	Far North District	3.8	3.5	Thames-Coromandel District	3.5	5.9
	Southland District	4.2	3.7	Tauranga District	3.3	1.8	Tauranga District	3.5	1.6
	Timaru District	3.4	2.6	Southland District	3.3	3.2	Far North District	2.8	2.5
		51.6			**60.2**			**68.6**	
	Christchurch City	3.3	0.4	Dunedin City	3.1	1.0	Rodney District	2.1	1.0
	Whangarei District	3.1	2.0	Rodney District	2.7	1.7	Auckland City	2.1	0.2
	Rodney District	3.1	2.9	Auckland City	2.5	0.3	Southland District	1.9	2.0
	Tauranga District	2.9	1.9	Gisborne District	2.2	2.0	Manukau City	1.6	0.2
C20	Gisborne District	2.8	2.0	Whangarei District	2.0	1.2	Waitakere City	1.5	0.3
	North Shore City	2.3	0.5	Chatham Islands District	1.9	71.2	Gisborne District	1.5	1.5
	Manukau City	1.9	0.3	Kaikoura District	1.8	20.1	Whangarei District	1.3	0.8
	Napier City	1.9	1.2	Wanganui District	1.5	1.3	Chatham Islands District	1.3	58.9
	Kaikoura District	1.8	18.5	Waitakere City	1.4	0.3	Grey District	1.3	3.6
	Chatham Islands District	1.6	64.3	North Shore City	1.4	0.3	North Shore City	1.1	0.2
		76.2			**80.7**			**84.2**	
	The rest	23.8	—	The rest	19.3	—	The rest	15.8	—
	Total New Zealand	**100.0**	—	**Total New Zealand**	**100.0**	—	**Total New Zealand**	**100.0**	—

Source: Statistics NZ, census usual residence areas and work status by industry 1981–2001.

There has been a decline in employment in the seafood sector in Northland and Gisborne regions. Concentration has led to a clustering of economic benefits in key regions (Nelson, Marlborough and Tasman) and to a lesser extent the port centres of the east coast of the South Island (Dunedin, Oamaru, Timaru and Christchurch). Re-localisation and restructuring in fisheries dependent areas has led to local and, O'Connor (1994) argues, most notably Maori employment impacts in Northland and the East Cape of the North Island. The QMS has realigned the industry and appears to have encouraged the development as well as spatial concentration of the processing sector.

Individual Firm Strategies

Re-regulation of the fisheries and the QMS led to new rounds of investment, but also restricted access to fish, forcing rationalisation. In response firms adopt a number of strategies. Broadly speaking these strategies can be summed up as attempts to grow more product and/or grow more value. Enterprises could either get more fish, strive to extract greater returns from the resources they had or do both. These strategies were informed by two factors. First, the QMS constrains the ability to expand effort. Second, the culture and history of individual firms.

Fishing firms have expanded horizontally and sought to amalgamate quota and ownership and achieve economies of scale, and, through diversification of quota holdings, economies of scope. Enterprises have grown more value developing value-added products and expanding vertically up and down the aqua-commodity chain. They have developed 'value-fishing' which hinges on rapid delivery to distant markets in order to take advantage of the premium that fresh fish receives (Hackett *et al.* 2005). Recognition that freshness receives a greater premium has led companies to develop local markets eschewing the transport costs of reaching overseas markets for local markets.

Firms have expanded vertically through the commodity chain. Expanding vertically may be construed as both a means of growing more products and growing more value. For companies such as Sanford and Sealord[3] vertical integration offers a vista of innovation that extends from catching and processing through to marketing, distribution and retail. By controlling product from 'seafloor to dining table' the opportunity to develop new and/or by-products from waste materials is enhanced. By vertically integrating companies are able to garner a greater share of the value of the products that are sold to consumers.

Fishing firms have moved beyond the purview of the QMS. Firms have shifted beyond the scope of New Zealand fisheries legislation and developed overseas fisheries resources, including international fisheries resources that fall outside the scope of national jurisdictions. Moreover, they have invested in aquaculture. Aquaculture has three benefits. First, it lies beyond the QMS. Production is largely a function of the space controlled. Second, there is certainty of production; barring environmental disasters the operator knows how much will be produced (Anderson 2002). Products can be supplied to the specifications of individual retailers. Third, aquaculture provides a security of rights that has yet to be eroded in the eyes of New Zealand rights holders.

---

3 The two largest seafood companies in New Zealand.

A number of small 'independent fishers' have exited the industry. Those that have remained have developed new skills and ways of operating. For example, Guard Fisheries, a small family owned fishing company, have developed local markets that hinge on rapid delivery of fish to its retail shops. Some fishers also align themselves to larger fish processors through formal and informal relationships to secure access to quota and fish stocks.

The QMS has displaced enterprise activity into other spaces and places where they may not be sustainable. Sanford and Sealord (amongst others) have developed overseas fisheries. Using skills developed in New Zealand deep-water fisheries and tacit knowledge of the practicalities of rights-based fishing, these companies have exploited overseas fisheries in ways that may not prove sustainable. They have moved quickly to exploit fisheries in less developed countries then exited as stocks decline and stricter regulations are introduced. More recently, Sealord has developed a 'southern hemisphere strategy' whereby it can supply hoki from a number of fisheries – all marketed under the Sealord label but beyond the purview of the QMS. Talley's Fisheries is the only large player to diversify beyond the seafood industry. It has built on core competencies in processing, and its location to expand into vegetable processing and then dairy products. More recently, building on this expertise developed in primary product processing, Talley's has entered the meat processing industry.

Individual firms respond remarkably quickly to changes in the regulatory regime. Firms have retired capital and effort from fisheries in the form of vessels, fish processing plants and labour. In addition, they have secured quota for vessels and processing facilities that cannot be retired. There are two distinct behaviours by firms that operate in New Zealand and overseas. In the New Zealand context these firms strive to build capacity and seek to become respectable, professional enterprises that are grounded in communities (Talley's), have addressed consumer and shareholder perceptions of sustainable development (Sanford), and fostered strong links with education and research providers (Sealord). In less rigorous regulatory environments they have tended to fall back into old habits of racing to fish and exploiting fisheries as quickly as possible. The QMS appears to have shifted aggressive growth-oriented activities into aquaculture and foreign fisheries, particularly those regulatory environments less stringent than New Zealand's.

Collective Firm Strategies

Fishing enterprises adopt collective behaviour in order to secure individual sustainable growth by collectively negotiating sustainability and profitability. Inter-firm cooperative behaviour under the auspices of the Nelson-Marlborough Seafood Cluster (NMSFC), Commercial Stakeholder Organisations (CSOs) and the Seafood Industry Council (SeaFIC) has, in conjunction with regulatory change and the QMS, enabled the fishing industry to renegotiate the debate on marine management in New Zealand.

Through SeaFIC, quota owners, particularly large quota owners, actively contest, engage and negotiate policy concerning fisheries. SeaFIC has scaled up the political voice of the industry. Regardless of the rhetoric concerning sound fisheries management practices and policies, its primary focus is on shaping the debate surrounding sustainable fisheries into one that best suits its members. The NMSFC is driven by tertiary education institutes with support and funding by the local Economic Development Agency and New Zealand Trade & Enterprise.[4] Rather than trying to attract new buyers and new customers the cluster works to develop skills, training, new products and innovative practices that add value to products that companies already produce. In this sense, for industry, the cluster is a means to secure resources: in this case skilled labour, new technologies and product innovations, but not fish. As such the NMSFC should be viewed as part of broader projects focused on economic growth and, for the State, regional development and regional growth. CSOs are an example of collective behaviour by quota owners. They represent Townsend's (2002) corporate bargaining. Some of them, but not all, may also represent privatisation of governance of the resource. CSOs are concerned with collective behaviour but only as it directly relates to the preservation and/or enhancement of property rights and associated value for quota owners and not necessarily fishers (Mansfield 2004a). Despite being bound within a rhetoric of co-management they are limited to quota owners. They are not democratic, egalitarian nor fair. Through CSOs quota owners strive to insert themselves into regulation, science and research. CSOs are a new space of governing. CSOs co-opt other actors and stakeholders into the process of fisheries management under the auspices of collaborative institutions, but only in the terms of the CSOs. Through CSOs, inter-firm cooperative behaviour is extended into the web of management, science and lobbying relationships that surround fisheries.

Discussion

The QMS has worked after a fashion. There is still a commercial fishing industry but the fishing industry is not the same one that fished prior to the introduction of the QMS. Some companies have survived and grown, many others, mostly independent fishers, have fallen by the wayside. Arguably the re-regulation of fisheries has proven economically equitable as quota markets treat all firms the same. However, the QMS is not necessary equitable. The costs of participating in the market are a greater burden for small independent fishers. Compliance costs fall disproportionately on smaller operators and fishers who lease quota. Independent fishers find it much harder to use quota as equity with which to borrow against and often lack the knowledge and skills to negotiate financial markets (Rees 2006). The QMS has enabled new rounds of investment in technology, investment overseas and outside the QMS including aquaculture and new ventures beyond fisheries by fishing enterprises. This investment was initially stimulated by the removal of part-time fishermen (with little or no compensation), the access to deepwater species previously fished by overseas companies and by a wider economic environment that

---

4 New Zealand Trade & Enterprise is the New Zealand government's national economic development agency.

encouraged many small fishermen to sell quota to larger firms because they had not recognised the true value of ITQ.

Firms have developed complex arrangements that secure continued access to fishing resources. The QMS provides an environment in which fishing enterprises work together. The institution of property rights removes the race to fish and has provided the impetus for firms to self-organise, acting in unison to preserve 'their' fishing rights that extended beyond the parochial, often fractured, organisations that existed prior to the QMS.

Commercial stakeholders have tried to shape interpretations of sustainable management. In one respect this should be construed as a strategy for securing sustainable growth. By constraining the debate on fisheries resources and how many fish are available or what the environmental effects of fishing are fishing companies are able to continue fishing. Through collective organisation, individual firms attempt to reshape what constitutes sustainable utilisation of fisheries resources (Massey and Rees 2004). This emergent corporatised governmentality curbs debate on sustainable utilisation through an ongoing *homo oeconomicus* discourse that includes political and ideological projects that are presented as objective, natural, technical truisms. The commercial lexicon surrounding sustainability acts as a particular vocabulary of motives through which power is exercised and dissenting alternative narratives are silenced.

That the QMS and ITQ results in firms' economic growth is debatable. Yes, firms are adding value and seeking to extract more value from the resources they have. Yes, firms are growing more products through aquaculture. However, increased productivity and performance has largely been on the back of two deep-water species – hoki and orange roughy. The near-collapse of orange roughy and recent substantial quota cuts in the hoki fishery will place pressure on larger fishing companies such as Talley's, Sealord and Sanford which all have significant fixed capital in deep-water fisheries.

Where quota is leased out by large companies or fish processors it is done so with the clear expectation that catch will be sold back to the processor regardless of better markets elsewhere. This coupled with the trend of small quota owners to remain in the business of fishing by leasing out quota when they retire rather than selling it has led to quota owners being distanced from human-environment relations and day-to-day practicalities of fishing. The distancing of ownership from use reduces fish and fishing to a simple accounting procedure. The QMS may need to be readjusted so that rights holders and rights users remain one and the same.

Firms need growth, development and profits. Firms such as Sanford, Sealord and Talley's have secured quota through purchase, acquisition and merger. Firms have developed the capacity to process and add value – growing more value to varying degrees. New processing technologies have resulted in increased labour efficiency and yield and also opened up the opportunity to develop products for specific markets. Skilled labour presents similar opportunities and the industry has made a concerted effort to develop a skilled labour pool.

Seafood firms face a tension between the constraints of fish resources (there are only so many fish in the sea) and public expectations that they will be competitive, prosperous and responsible. They have a number of options for securing access to resources (Figure 8.1a). They can continue to fish in New Zealand waters, in which case they can fish inshore or in deep-water fisheries. Firms may invest in aquaculture within New Zealand waters. Although this has been stymied by recent ongoing debates concerning access and a permit moratorium. Firms may fish overseas. They can fish in international waters or they can fish in another country's EEZ, usually through joint ventures. Each of these strategies involves costs and benefits. Generally, the costs have proven to outweigh the benefits for many smaller firms and they have chosen to exit the industry.

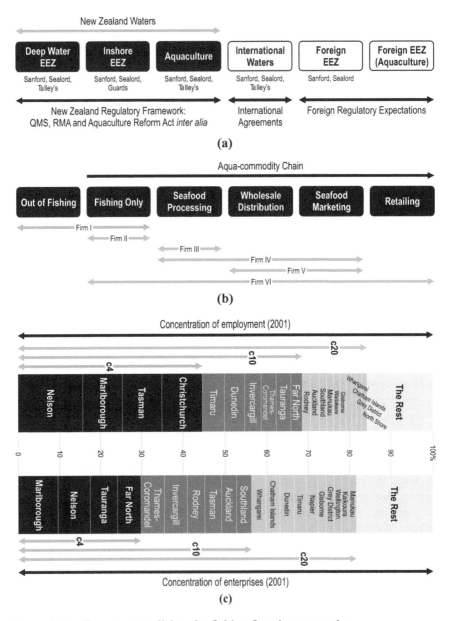

Figure 8.1   **Re-conceptualising the fishing firm in geography**

Seafood firms may continue to be competitive and prosperous by changing the locus of activities (Figure 8.1b). Such a change may shift firm activities beyond the act of fishing. Firms that fish can shift along a continuum between fishing and being completely out of fisheries. For some firms, as core competencies develop, the emphasis shifts to activities further along the commodity chain. For others, they exit fishing and the seafood sector altogether.

Shifting activities beyond fishing and at the most extreme out of fisheries all together calls into question the very nature of what exactly constitutes a fishing company the ways in which we view New Zealand's seafood sector. The drive to extract value results in new competencies and skills, leading to new possibilities. As new possibilities are presented and firms find new opportunities the attraction of fishing may decline. For many firms the challenges of sustainable development have been resolved through industrialisation. What is clear is that for most of the large firms and some small operators (such as Guards) it is fallacious to view them as solely fishing firms (Figure 8.1b).

The spatiality of production is allied to securing resources, the ways in which value added is constructed and the very nature of the firms involved (Figure 8.1c). The nature of New Zealand fishing has led to the concentration of employment and to a lesser extent enterprises in the top of the South Island. Firms concentrate processing and ports servicing of vessels in one site to achieve economies of scale and scope. Coupled with diversification into aquaculture and processing, the top of the South Island represents a locus for economies of scale and scope. Continued agglomeration in the top four regions raises questions as to the very nature of fishing and the firms that continue to fish beyond these regions. In addition, as firms shift away from fishing in New Zealand and extend further along the commodity chain (Figure 8.1a and 8.1b), opportunities to shift processing and value-added to spaces beyond New Zealand regulation and away from New Zealand labour will present challenges for the communities that depend on processing facilities for their livelihood.

Conclusion

In New Zealand responses to QMS and property rights reform have been persistently diverse. This diversity arises out of the polycentric multi-scalar nature of the fishing industry. The QMS has imposed constraints on fishing enterprises which responded in various ways. Arguably, a substantial move towards sustainable management of stocks occurred. This sustainable management has involved both an increased fisheries science effort by the state and industry, and restricted TACC. This has led to a realignment of the fishing industry. First, overseas where less than exemplary behaviour confirms the need for state regulation missing in other jurisdictions. Second, firms have expanded into aquaculture. However this local expansion of effort is constrained by delays with seabed and foreshore legislation. Third, there has been 'competition' among companies. The cooperative aspect is the collective use of shared knowledge to pursue common benefits. Competitive aspects refer to the ways in which this shared knowledge is used to achieve private benefits and out-perform others. Quota owners and specifically larger firms have organised collective behaviour to preserve property rights, to bid for science and to gain access to resources other than fish. However, firms still compete against each other. Fourth, as TACC has declined in key stocks there is ongoing rationalisation of plant and processing capacity. Fifth, there have been 'discoveries' of new stocks. Notwithstanding the importance of these stocks to large fishing companies, these stocks, such as southern blue whiting and ling, are not so much new as newly exploited by New Zealand companies.

Peter Talley (1999, 252) has argued that if we are to be serious about the business of fishing we need a lot more economic advice and a lot less biological opinion. Previously I have dismissed this as a thinly veiled attack on fisheries science, and implicitly, the Ministry of Fisheries (Massey and Rees 2004). Here, rather, I would like to suggest that in one sense Talley is correct. The increasingly co-constructed regulatory context has forced firms that fish to develop new ways of doing fisheries that extend beyond the 'simple' act of fishing. Economic advice, and not just neoclassical modelling, on precisely how new relationships should be formed, what ways in which firms could work together and how they should behave in globalising fish processing and fish product markets may prove essential for the continued success of individual firms and the industry as a whole. We should be less inclined to represent fishing companies as unscrupulous and conscienceless, and rather as complex, at times contradictory, entities containing actors who are situated in complex social, economic and in the case of Maori fishing companies cultural relationships.

Prior to the introduction of the QMS the New Zealand seafood industry focused on fishing, filleting and flogging fish. Little attention was paid to value-adding or establishing relationships beyond those with wholesalers and distributors overseas. Post QMS analysis of the industry demonstrates that many participants have altered their behaviour. There have been a variety of strategies, some successful and others less so, but all illustrate that in order to remain competitive and prosperous firms that fish have had to move beyond the 'simple' act of fishing New Zealand waters. Firms are moving up, and down, aqua-commodity chains, repositioning and reconfiguring themselves as they strive to compete. Furthermore, these commodity chains operate at increasingly extensive geographic scales. That is to say, firms operate in a number of territorially defined political-economic systems and scales. For example fishing firms operate in local communities (such as Nelson) at regional and national scales (as in the case of companies fishing hoki) and at supra-national levels (such as Sealord initiating a 'Southern hemisphere fishery'). Furthermore, operating in these interrelated scales of the local, the national and the global firms are subjected to interrelated scales of regulation.

References

Anderson, J. (2002), 'Aquaculture and the future: why fisheries economists should care', *Marine Resource Economics*, vol. 17, pp. 133–51.

Anderson, L. (2002), *Fisheries Economics: Collected Essays*, Ashgate, Burlington.

Arbo, P. and Hersoug, B. (1997), 'The globalization of the fishing industry and the case of Finnmark', *Marine Policy*, vol. 21, pp. 121–42.

Connor, R. (2001), 'Changes in fleet capacity and ownership of harvesting rights in New Zealand Fisheries', in Shotton, R. (ed.) *Case Studies on the Effects of Transferable Fishing Rights on Fleet Capacity and Concentration of Quota Ownership*, FAO, Rome, pp. 151–87.

Dayton, P., Thrush, S. and Coleman, F. (2002), *Ecological Effects of Fishing in Marine Ecosystems of the United States*, Pew Oceans Commission, Arlington, Virginia.

Fløysand, A. and Lindkvist, K.B. (2001), 'Globalisation, local capitalism and fishery communities in change', *Marine Policy*, vol. 25, pp. 113–21.

Hackett, S., Krachey, M., Brown, S. and Hankin, D. (2005), 'Derby fisheries, individual quotas, and transition in the fish processing industry', *Marine Resource Economics*, vol. 19, pp. 45–58.

Hersoug, B. (2002), *Unfinished Business: New Zealand's Experience with Rights Based Fisheries Management*, Eburon, Netherlands.

Hutchings, J. (1999), 'The biological collapse of Newfoundland's northern cod', in Newell, D. and Ommer, R. (eds) *Fishing Places, Fishing People: Traditions and Issues in Canadian Small-Scale Fisheries*, University of Toronto Press, Toronto, pp. 260–75.

Ludwig, D., Hilborn, R. and Walters, C. (1994), 'Uncertainty, resource exploitation and conservation: lessons from history', *Ecological Applications*, vol. 3, pp. 547–9.

MacFarlane, A. (2002), Editorial, *Seafood New Zealand*, vol. 10, pp. 4–6.

Mansfield, B. (2001a), 'Property regime or development policy? Explaining growth in the US Pacific groundfish fishery', *Professional Geographer*, vol. 53, pp. 384–97.

Mansfield, B. (2001b), 'Thinking through scale: the role of state governance in globalising North Pacific fisheries', *Environment and Planning A*, vol. 33, pp. 1807–27.

Mansfield, B. (2004a), 'Rules of privatisation: contradictions in neoliberal regulation of North Pacific fisheries', *Annals of the Association of American Geographers*, vol. 94, pp. 565–84.

Mansfield, B. (2004b), 'Neoliberalism in the oceans: "rationalisation," property rights, and the commons question', *Geoforum*, vol. 35, pp. 313–26.

Massey, E. and Rees, E. (2004), 'Sustainable utilisation of fisheries as governmentality: constraining the potential for ecosystems-based management', *New Zealand Geographer*, vol. 60, pp. 25–35.

Newell, D. and Ommer, R. (1999), *Fishing Places, Fishing People. Traditions and Issues in Canadian Small-Scale Fisheries*, University of Toronto Press, Toronto.

O'Connor, M. (1994), 'On the misadventures of capitalist nature', in O'Connor, M. (ed.) *Is Capitalism Sustainable? Political Economy and the Politics of Ecology*, Guilford Press, New York, pp. 125–51.

Rees, E. (2006), In What Sense a Fisheries Problem? Negotiating Sustainable Growth in New Zealand Fisheries, Unpublished PhD Thesis, The University of Auckland, Auckland.

Statistics New Zealand (1981–2001), The Census of Population and Dwellings (2001 Areas), Prepared for The University of Auckland, Usual Residence Areas and Work Status by Industry for the Employed Census Usually Resident Population Count, Aged 15 Years and Over, Statistics New Zealand, Wellington.

Statistics New Zealand (1984–2002), *Key Statistics*, Statistics New Zealand, Wellington.

Straker, G., Kerr, S. and Hendry, J. (2002), *A Regulatory History of New Zealand's Quota Management System*, Economic and Public Policy Research Trust, Wellington.

Talley, P. (1999), Fishing Rights – an Industry Perspective. ITQs and Fishermen's Attitudes: The Change from Hunter to Farmer, FishRights99, Fremantle, Western Australia.

Townsend, R.E. (2002), *The Role of Coasian Bargaining in Efficient Fisheries Management*, Eleventh biennial conference of the International Institute of Fisheries Economics and Trade (2002), Fisheries in the global economy, Wellington, New Zealand.

Wright, C.S. (2004), *Fisheries Management: Unhelpful, Unkind, and Unjust*, Twelfth Biennial Conference of the International Institute of Fisheries Economics and Trade 2004: What are Responsible Fisheries?, Tokyo, Japan.

Chapter 9

Constructing Economic Objects of Governance: The New Zealand Wine Industry

Nicolas Lewis

Introduction

Key theoretical debates in political and economic geography have highlighted the significance of relationality and the constitution of *economic objects* and *spaces of governance* (Peck 2004; Yeung 2005; Gibson-Graham 2006). The hard-wired relational categories of political economy have come under challenge from a language of new 'state spaces' or 'economic spaces' (Leyshon *et al.* 2003; Peck 2002). This language offers opportunities to develop post-structural political economies that challenge mainstream economic geography (Le Heron 2006). In such terms, spaces of governance are configurations of economic relations and activities, politically inspired representations of them, and institutions of governance; and scale is a construct (Sheppard and McMaster 2004). They are not pre-formed but coalesce and are shaped into assemblages such as 'the economy' – a construct of social, political and economic processes in the mid-twentieth century (Mitchell 2002). Apparently foundational categories are constructs for knowing and managing economic and social relations. They are produced discursively and by a 'politics of calculation' and associated practices, measurements, and boundary setting. In this chapter, I focus attention on the ways in which industry is being reworked in these terms.

This chapter builds on Jessop's (1997, 58) interpretation of governance as a term that refers to all forms of coordination of economic relations that involve neither market forces nor formal hierarchies. I draw attention to both inter-enterprise relations and non-market coordination encompassed in notions of industry. Industry is widely invoked as a foundational category in political economy to relate activities linked functionally to the production of a particular commodity. It is also often treated in political discourses as a subject with a particular identity, reason and capacity to act. This chapter denaturalises these taken for granted applications to interpret industry as an analytical and theoretical construct – an assemblage of social and economic relations, political rationales, expert discourses, institutions, and related agents, through which diverse social forces, projects and organisational capacities are mobilised (Lewis 2005). Working with data gathered from two research projects (over 100 interviews with wine enterprises and extensive statistical records collated from industry sources – see Moran 2000; Lewis 2004), I use the case of the New

Zealand wine industry (NZWI) to argue that industry has emerged as a key space of neoliberal governance. I begin by developing the notion of industry as a space of governance from a more conventional stance, map the institutions of governance at work in the New Zealand wine *filière*, and go on to describe the deployment of the NZWI as a space of governance in contemporary political programmes.

Industry as Space of Governance

Economic geographers have deployed notions of commodity chains, production networks, untraded interdependencies, learning regions, and institutional thickness to examine the embeddedness of decision making by economic enterprises. Whether acknowledged directly or not, the notion of *governance* (the economically conditioned and socially constructed arrangements, organisations, rules and processes that coordinate relations) is central to this work. Multiple mechanisms of governance are involved in the interplay of corporate hierarchies, industry organisations, state regulation, social institutions, or social networks (Hollingsworth and Boyer 1997). Lewis and colleagues (2002) argue that in these terms, enterprises are embedded in three spheres of governance: inter-enterprise relations, socio-spatial relations, and national regulatory structures. Each sphere is associated with a particular body of knowledge that privileges different mechanisms of coordination and control, agency, and points of entry to understanding political economy (Table 9.1).

Table 9.1   Elements of a theory of governance

Literature	Discipline	Point of Entry	Agency	Conditioning Process	Sphere of Governance
New institutional Economics	Economics	Boundaries of the firm	Enterprise	Business	Enterprise
Institutional Thickness	Regional Studies	Social institutions	People in place	Socio-spatial reproduction	Territorial
Regulation Approach	Political Economy	Macro social order	The state	Policy	Reglementary

Source: After Lewis *et al.* 2002.

Each sphere of governance is interpenetrated by institutions that operate in the other spheres and are mutually constitutive. In articulation, they comprise spaces of governance. Lewis and colleagues (2002), working with the French notion of the *filière*, deploy Herrigel's (1994) concept of an *industrial order* to edge towards an interpretation of industry in these terms.[1] They highlight interrelations among

---

1 An institutionally self-defined and self-governing, system of production that coheres around the interrelationships that comprise the *filière* and the apparatuses of governance that coordinates them.

economic enterprises that give form to commodity chains or production networks, the interdependencies that raise governance issues, and the mechanisms of governance that have arisen. They observe that economic geographers, and popular commentators on industries alike, tend to ignore the problematic boundaries, representational force, properties of identification, and power relations of industry as governance. This chapter expands on this observation.

Understood as industrial order, industry is constituted by an interplay of work practices, social and economic interdependencies, path dependencies, and shared experiences and ideas. A meaningful configuration of relations, institutions, understandings, authorised knowledge and practices, and shared beliefs about its character, it takes on an identity and a spatiality inhering in place, path dependencies, and the reach of different institutions as well as the materiality of flows and relations (see Storper 1997). This chapter argues that the notion of industrial order can be further enhanced by drawing on post-structural ideas of the discursive constitution of political-economy categories and deeper understandings of identity as politically constituted and connected to the interplay of power and knowledge and its governmental and subjectifying forces.

Governance, Value Chains and *Filières* in Wine

In land-based production, the territoriality of an industrial order is more pronounced. It encompasses the significance of specific knowledge of local physical environments, dense social and professional networks, and a history of interdependent relations among enterprises. In wine, these factors are particularly significant, adding complexity to what is a straightforward production circuit from the growing of the grapes, to the making of wine, bottling and packaging, distribution, and sale. A highly differentiated and high value product, wine has a complex territoriality of production (Moran 1993a). In production, this territoriality emerges from a place-specific articulation of production environment and technique. Often captured in the French notion of *terroir*, a key concept in the global culture of the vine (Vaudour 2002; Moran 2006),[2] it involves place-conditioned processes of learning, development trajectories, and associations of quality. Place conditions growing conditions, decisions in the vineyard, and the craft of the winemaker, and generates an extensive range in quality differentials. Place labelling of bottles, place centred reputations, and 'placed' stories about particular wines are crucial to market success in heavily differentiated markets. Contemporary, more active branding practices link wider imaginaries of place and the reputations of particular winemakers to these configurations (Lewis and Kelly 2006). Quality configurations of producer, aspiration, variety, and style are overlain by regional association, and have been branded as such for centuries.

---

2 *Terroir* has both a narrow meaning that equates it to soil, and a broader interpretation similar to that of the English word territory that refers to the natural and human characteristics of a delimited area of land. It appears in either or both forms in the circulation of knowledge that links the understanding and associated practices of producers, distributors, wine writers and consuming public to notions of quality (Moran 2001).

Quality wine *filières* are thus marked by complex untraded interdependencies and widely recognised collective interests. In many settings, these give rise to webs of institutions of shared winegrowing and making traditions and practices at the local level, social conventions to do with managing collective interests and reputations, and laws to do with representing place in the product (Moran 2000; Barker 2004). The most well known of these webs is the AOC, which subjects much of French winemaking to tight regulatory scrutiny, specific production rules, and a raft of local and regional conventions (Gade 2004; Moran 1993b). These are woven into a rich set of placed social practices, norms and conventions, and are continually contested and renegotiated through these strands and their weavings and the changing positions and trajectories of enterprises and region.

New Zealand Wine Spaces

New Zealand is a small producer of wine by global standards (less than 30,000 ha or 0.05 percent of global production), but production has grown rapidly in recent years (Lewis 2004). Although less rigidly and finely contoured than the French model (by reglementation and centuries of history), the institutional landscape of wine production in New Zealand remains thicker and more complex than commonly held (Barker *et al.* 2001). A mix of territoriality, multiple apparatuses of governance, enterprise forms, small size, very rapid growth, and ethos of cooperation have produced a particular path dependency, strong industry identity and a distinctive industrial order.

New Zealand wines are characterised by their high quality and high value on international markets. Collectively, they achieve price premiums in international markets, favourable reviews from wine critics, and success in international wine competitions. Production is marked by regional diversity, regionalised development trajectories, increasing regional specialisation, and a strong sense of regional identity (Moran 2000). It is also marked by an increasing diversity of enterprises along several axes – size, age, ownership structure, resource bases, business practices, and extent of vertical and horizontal integration. Other defining features include increasing export dependence, the success of which is dependent on sustaining New Zealand's reputation for quality production and the price premium its wines enjoy on international markets (Banks *et al.* 2007). The cooperative behaviour and industry level institutions that have played a prominent role in building and sustaining this reputation for quality remain crucial (Lewis 2003).

Figure 9.1 maps these apparatuses and institutions of governance onto the frame suggested by Lewis and colleagues (2002). Key features of this mapping include contract grape-growing, the emergence of processing contracts, local trust-based relations and collective learning, overlapping professional and social networks, informal knowledge sharing, formal industry organisations, accepted norms of conduct, voluntary regulatory arrangements, and a raft of industry-specific regulation. Each apparatus is located to emphasise its work in one sphere of governance and the primary agency working through them – the firm, people in place, and the state. Each encompasses elements of the other spheres, emphasising their mutual constitution.

As an economic institution, for example, the family is both social institution and capitalist enterprise, and is deeply rooted in place (Moran *et al.* 1993); New Zealand Winegrowers (NZW) is a national association of winemaking enterprises, but is sanctioned by the state, in part defined by legislation, and responsive to and mediated by regional/local concerns; and social networks are formed in all spheres and link state, local social institutions and inter-enterprise relations.

Figure 9.1 defines a space of governance within which apparatuses of governance are positioned according to the source of their form, features and authority – state, place or firm. Particular spheres of governance are defined by the positioning and relationality of these apparatuses in terms of the articulation of different sources of governance. These spheres are captured in the sectors marked out by grey lines as territorial, sectoral and reglementary. These spheres are neither discrete nor tightly bounded. Rather, they are loose articulations of state, firm and place influences and configurations of apparatuses that condition actor and activities that exist simultaneously in all three spheres. The national boundary indicated by the heavy

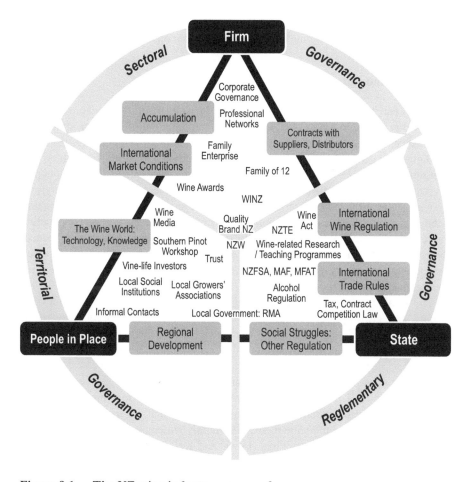

Figure 9.1   The NZ wine industry – space of governance

dark line is similarly porous, and actors and actions that appear of the national space are also part of the international and subject to external forces routinely brought to bear on the national space of governance. Again these are positioned to suggest the points of relationality at which these forces penetrate the national industry.

Figure 9.1 is drawn for a specific moment in the development trajectories of industry, enterprises, and a particular territory (New Zealand). It delimits a particular configuration of governance and a particular segment of one *filière*. Each of these focuses invokes artificial boundaries – temporal, spatial and sectoral. That is, the industrial order presented is dominated by the recent history of rapid expansion. Rapid growth has highlighted certain tensions (access to land, materials and knowledge), but suppressed others such as those among winemakers over market share and between grape growers and winemakers over prices – one reason for the centrality of collaborative governance and the industry organisations. The key point is that of contingency. *The axes of difference, which might in other times and places generate conflict have been suppressed by universal growth and narratives of success.* Figure 9.1 also recognises the openness of any industrial order exposed to change from both exogenous and endogenous forces – the life course of enterprises, new technologies, changing market conditions, and so on.

Enterprises and Enterprise Governance

Figure 9.1 can be interpreted by looking into it through any of its three apices to give analytical privilege to a particular sphere in the articulation of the three spheres of governance. Taking a view through the lens of economic governance, for example, highlights the governance arrangements that condition intra- and inter-enterprise relations (Table 9.2). Figure 9.1, however, recognises explicitly that economic governance also has both territorial and reglementary components. Decisions to do with vineyard location, varietal choice, vineyard management, winemaking technique, marketing, and expansion will be conditioned by national legislative frameworks, international trade rules to do with geographical indications, local contests over property rights and land use, knowledge produced by state research institutions, and locally generated and disseminated tacit/empirical knowledge of, for example, the physical qualities of different sites.

The small family enterprise has dominated the history of wine production – owning land, growing grapes, organising labour, making wine from its vineyards in its own winery, and emphasising the significance of the interdependency between grape growing and winemaking, technically and organisationally. Strong networks of trust, flexible work practices, reduced transaction and management costs, and deep knowledges of *terroir* give the family strong competitive advantages (see Friedman 1978). In wine, this integrated grape growing and winemaking family enterprise supports the enduring romantic, neo-artisanal images that are used extensively in marketing material and the wine media. The family winemaker is a central identity, even creation myth of wine cultures (Lewis and Kelly 2006), and a characteristic feature of the world's wine industries (Anderson 2003).

Table 9.2   Sectoral governance in the New Zealand wine *filière*

Enterprise Relations	Enterprise-level	Reglementary	Territorial
Intra-enterprise	Enterprise form	Industrial relations, company law, tax law	Family life cycle Local knowledge Local government
Inter-enterprise	Spot market Trust (non-formal) Contract Management	Contract law, competition law, trade policy Supranational mediation Industry specific legislation	Social institutions (trust) Proximity/shared interest Regional political economy
Multi-enterprise (sectoral)	Local associations National associations (WINZ and NZGGC)	State sanctions Industry specific legislation Competition state State research institutions	Cultures of place Proximity/shared interest Local associations Regional political economy
Sub/super-enterprise	Construction of knowledge	Economic and social legislation	Reproduction of place

In New Zealand the 480 small, mainly family-based, winemaking enterprises make up more than 90 percent of all winemaking firms, whilst the largest companies all have backgrounds as family enterprises.[3] Most of the 873 grape growers are also small, family-based operations.[4] In practice, family enterprises vary significantly in their form, size and strategies. The romantic imagery borrowed from the Old World of wine attaches most easily to the long-standing Dalmatian winemaking families of West Auckland. However, the category also encompasses the first/second generation family winemakers who have set up over the last 30 years; farming or orcharding families that have diversified into grapes; mid-large sized firms that have emerged from these origins, lifestyle investors in grape growing; and wealthy professionals or industrialists looking to invest discretionary capital in land and/or consume wine's romantic imagery and the rural idyll (often in partnership with new family enterprises).

As they have grown, family enterprises have devised new strategies, organisational forms, and coordinating institutions, fracturing the *filière* around a new complexity of inter-enterprise relations. Increased professionalisation, functional specialisation, new complexities of market access, problems of access to capital, and new opportunities in new regions have all contributed to a complex set of enterprise interrelations. Delegat's Wine Estate, New Zealand's third largest wine enterprise, for example, is still family controlled and New Zealand owned, but is listed on the secondary market, raised capital for its new winery on the securities

---

3 The fourth largest, Villa Maria, remains family owned and governed.

4 Some, however, have significant holdings while several are corporate growers or land investment firms.

market, and has used contract processors to make its wines and vehicles such as an Australian investment fund to develop its vineyards. Whilst some family enterprises have expanded to establish restaurants and gift shops, new entrants and growers with plans to develop their own brands may share plant or contract winemakers to make wine for their own labels. At the same time, large commercial processing enterprises now make wine for label owners (including British supermarkets, corporate growers, and freelance winemakers who own no vineyards) and specialist firms run labour gangs, act as viticultural consultants, and broker secondary wine markets or retail over the internet.

At the other end of the size spectrum, the six largest producers, each producing more than 2 million litres a year, together produce more than 75 percent of total production. Three are owned by transnational drinks companies,[5] whilst foreign interests in total control an estimated 50–60 percent of total production. The largest, Pernod Ricard New Zealand, makes 30 percent of all New Zealand wine. These companies share corporate forms with many of the 44 mid-sized companies (sales between 200,000 and 2 million litres) – some owned by private shareholders, some by other multinationals, and some by larger family interests.

Running through this increasing complexity (Lewis 2004), two sets of interdependencies and associated tensions have a particular significance in the constitution of an NZWI – those between grape grower and winemaker; and the shared interests of all participants in 'Brand New Zealand Wine'. Whilst contracts and trust-based arrangements of multiple forms facilitate many of the relations between growers and wineries, attention here is on multi-enterprise or sectoral relations. Enterprises have acted to build various forms of associational governance beyond vertical integration or contractual relations with other firms. There have been collective initiatives of various forms in marketing as well as production (the Family of Twelve, Cellars of Canterbury, Rapaura Vintners, Central Otago Wine Company, and Gisvin, among others). Such multi-enterprise relations are highlighted in the central portion of Figure 9.1, but are overshadowed by the significance of the regional and national winemaker and grower associations.

Associational Governance

Occupying significant positions within an industrial order (Coleman 1997), such subjects of associational governance become both subjects and objects of government (Perry 2001; Lewis 2003). They promote, regulate, and represent the industry, and may coordinate activities or undertake them directly. The literature suggests that such associations will be influential where there exists a history of cooperative relations, overlapping socio-spatial bonds, significant shared interests, and/or strong industry-government links (Streeck and Schmitter 1985; Coleman 2002). Each of these conditions holds in the case of New Zealand wine. The New Zealand Winegrowers (NZW), formed in 2002 from a merger of The Wine Institute of New Zealand (WINZ)

---

5 The capacity to add value beyond the winery has attracted major international brand owning drinks companies, which have become increasingly backward-linked into winemaking companies through distribution agreements, merger and takeover (Burch and Goss 1999).

and the New Zealand Grape Growers Council (NZGGC),[6] play central roles in the constitution of the NZWI. Membership of WINZ and NZGGC has been effectively compulsory by the levying of grapegrowers and licensing of winemakers (Lewis 2003). Growers belong to their local grape grower associations and through these to the NZGGC and now NZW, whilst wine makers belong directly to WINZ.

Historically, these organisations have exercised significant influence, particularly in holding together diverse interests (along lines of size, region, among other axes of difference and potential contest – social as well as commercial) (Lewis 2003). WINZ, formed in 1975 after developing a novel solution to size-related diversity by guaranteeing significant representation on its Board for smaller enterprises, was the more prominent of the two organisations. Its activities included building national industry institutions such as the Air New Zealand Wine Awards, the annual industry 'Bragato Conference' and Wine Exporters' Forum, the publication of an official industry journal, the establishment of sustainable winegrowing programme (SWNZ). Each consolidated industry networks and was supported by the open door policy adopted by the CEO of WINZ (Lewis 2003). Regional associations, particularly growers' associations, promoted regional interests through regional plans, educational workshops, lobbying, and market development.

WINZ (and now NZW) became accepted by government as the voice of the industry. Lobbying and regulatory liaison became central to its work. Concerns have ranged from excise, to geographical indications, international trade access, and research spending. WINZ lobbied heavily for the introduction of the Wine Act (see below) and legislation to enable the formation of the NZW, which now coordinates interdependencies among winemakers and grapegrowers. WINZ developed strong relationships with successive governments, and it is now widely cited as a model of the state working with a winner industry to develop research consortia, generate regional employment, and promote export markets. Trade promotion and marketing have dominated activities since the mid-1990s. Alongside economic development agencies, it developed, promoted, and acted as guardian of Brand New Zealand Wine and the 'Riches of a Clean Green Land' logo.

Industry organisations have successfully promoted collaboration and the sense of a singular and inclusive NZWI, expressed in the formation of NZW and widespread references by participants to 'the industry' as a sentient subject. They have worked to overcome ethics of independence, differences in enterprise sizes and aspirations, the structural opposition between growers and winemakers, tensions between new entrants and long-term participants, and a strong regionalism (Lewis 2003). At times, such as in the struggles in the late 1980s over grape prices and the institutions for establishing these and conventions for payment (Barker *et al.* 2001), these differences have been highly divisive. On-going tensions exist around setting of levies, identifying markets for generic marketing programmes, funding regional promotions, and fixing the generic-user pays ratios for trade fair costs, setting research agendas. They are joined by new tensions around sharing information with new entrants and fixing geographic indications (GIs). Some of these form along size

---

6 The two long-standing and dominant bodies of associational governance in New Zealand wine.

axes, and the assent of the largest companies must always be gained (which all have a voice on the Board). However, industry organisations have worked to suppress tensions by enhancing routine interactions among participants, cultivating shared experiences and knowledges, nurturing mutual interests in Brand New Zealand Wine, and capitalising on the collective momentum generated by the recent rapid growth. WINZ and now NZW have helped ensure that the larger companies have benefited from the award success and artisanal imagery of smaller firms, and that smaller firms have fed off the market penetration of the larger companies.

Place

Place is significant in the operation of many of the apparatuses recorded in Figure 9.1. As well as territorial regulations, place-based governance emerges from proximity and repeated contact, the local embeddedness of economic relations, the construction of collective understanding, and the generation of trust. Family enterprises are embedded in the local community, which conditions the ways in which they forge economic relations and engage with knowledge networks (Moran 2000). Place-based social relations range from cultural and community links among Dalmatian families to the frontier friendships and shared problem solving of the pioneering Central Otago winemakers. In wine, decision making is rooted in the generation and communication of local knowledge of local grape growing environments and associated winemaking/growing techniques. Such knowledge is built through social and professional exchange of empirical experience, as well as more formal consultancy arrangements, and research and education programmes. With production histories and experience of different environmental conditions relatively recent, knowledge flows around social networks are crucial. The same points hold for knowledge of markets, and the struggles to open new markets.

Local social and professional networks facilitate multiple localised exchanges and reinforce the collaborative endeavours and senses of industry fostered by winemaking and grape-growing organisations in their meetings, educational seminars, and newsletters. Labour supply, collective purchasing of inputs, irrigation initiatives, responses to disease, political lobbying, regional wine trails and festivals, regional marketing initiatives, and visits by international wine media are coordinated through these networks. Local industry organisations have generally emerged from the ground up, out of these networks, shared cultural backgrounds, and shared experiences (Cooper 1977; Barker *et al.* 2001). Alongside less formal social networks, these organisations mediate competitive tensions between and among growers and winemakers, small and large enterprises, and established businesses and new entrants.

In all regions, winemakers have regular, often informal, meetings where they taste and discuss wines and share their knowledge, outside or beyond the local industry organisations. These meetings extend from the chance meetings of living and working in the same place to more deliberate engagements to discuss particular issues such as those of the Southern Pinot Noir Workshop (Moran 2001) or the more recent Syrah Workshop in the North Island (Sundae, Pers. Comm. 2006). Joint wine and tourism promotional events are increasingly important exercises in collective marketing and tourism development. Other more individualised, contractualised and exclusive local associations, such as the collective use of mechanical inputs or winemaking facilities or cooperative arrangements in wineries or marketing initiatives (Lewis and Kelly 2006), remain conceived in localised professional networks and relations of trust. In whatever form, these encounters become factors in decision making and help to coordinate decisions.

The State

Wine enterprises are subject to the full range of general legislation that impacts on enterprises – land use and resource management regulations; monetary and fiscal policy; company, tax, contracts, and industrial relations law; and overseas investment regulations. Trade policy, food security, bio-security provisions, and truth-in-labelling have a particular resonance in a globalising food and alcohol industry. The state has also had specific impact on the development of winegrowing and the constitution of the NZWI. The Wine Industry Development Plan of 1981 drove the early 1980s growth in production for domestic consumption, which resulted in a wine glut and intense competition for market share. Government later initiated a formative vine replacement scheme in 1985 which sought to cut production. Over 25 percent of vines were removed, predominantly those of lesser varieties and those grown with poorer viticulture (WINZ 1986), allowing growers to take advantage of technical developments in viticulture and the emerging emphasis on quality by replanting with improved blends of root-stock, clone and variety on altered trellising systems (Barker *et al.* 2001). Further boosts of one form or another were driven by later deregulation of the wine trade – international trade, sales from supermarkets, extended licensing hours, Sunday sales and so on.

Wine specific regulatory intervention has also had formative effects. Levies on grape growers and licensing rules for winemakers have effectively regulated for compulsory membership of WINZ, NZGCC and NZW. Government agencies have worked alongside WINZ in testing, certifying and promoting exports, funding research and education, and reinforcing industry identities and cooperative practices. NZW continues to liaise closely with economic development agency NZTE on market development, and trade agency MFAT on market access. Alongside MFAT it has negotiated a Mutual Acceptance Agreement on wine making practices with fellow new world wine producers (see Barker 2004). NZW and regional associations have also worked closely with government in the specification of GIs for acceptance by Europe. In 2002, the passage of the Wine Act brought all wine related regulation to do with production, sales, excise, health and safety and so on under the same legislation. The Wine Act works through a Wine Standards Management Plan to impose health and safety and environmental standards, various, codes of practice, schedules of best practice, HACCP analysis, audit techniques, and labeling to place industry self-regulation and NZW at the centre of regulation. It is administered in close association with the NZW, the formation and functions of which are facilitated and mediated by The Act. With the Wine Act and in all these other ways, the state mediates external influences, works alongside NZW, and has been an instrumental in building an industrial order – with NZW at the centre.

Identity and Discourse

The associations of wine with nation are infused by a discourse of quality production. Images of quality can be readily attached to wine, which is a paradigmatic example of the possibilities inherent in the turn to quality in governance (Morgan *et al.* 2006). Its dominant discourse of *terroir* expresses creativity in production, artisanship in tradition, social/cultural prestige, product differentiation, and rarity values. It suggests a quality gradient. Elite wines have a high value, and are marketed through place signs, which add significant brand values (Beverland 2005). The aesthetics of many of the production settings in New Zealand add new dimensions to these associations, and attach added returns through tourism. Place mediates associations between environment, quality, social prestige, wine knowledge, and economic values, and generates collective rents, whilst GIs and labeling rules govern these relationships and their expression.

In New Zealand these associations are crafted into a set of understandings and practices that establish 'Brand New Zealand Wine'. Although a far more generic set of associations of place with product, this brand is commonly represented by the logo 'Riches of a Clean Green Land', which defines New Zealand wine by reference to key environmental and cultural images – rugged, wide open, green, and land-centrism. These images echo the romantic imaginary of the artisan family enterprise and *terroir*, and reinforce reputations for high quality. Quality is seen by participants, wine media and government alike as identity defining, and the reputation for quality as pivotal to the future success of all enterprises.

The discourse of quality is also industry making. Quality imaginaries inform winegrowing and marketing practices, link the brand-led strategies of large firms to the efforts of smaller enterprises to exploit artisanal imagery, and highlight the collective interests of all wine enterprises (from grape growers to winemakers, distributors, labour contractors, and wine writers). As well as drawing from and reinforcing the sense of collective ownership of the recent growth and successes, quality imaginaries also mediate the practices of diverse enterprises and inform production and marketing knowledges. They relate the practices, interdependencies, and governance apparatuses recorded in Figure 9.1 to industry identity. This understanding of 'quality' resonates with the concerns of writers announcing the quality turn (see Morgan *et al.* 2006), but is a more powerful notion than fetishised accounts of the local that stress artisanship, rurality, smallness local embeddedness, and the local product. It connects representations of place to production and marketing imagery, the growth of brands, and contests over value added in agro-food complexes. Far from a romantic notion of identity set up in opposition to 'globalisation', this notion of quality gives winegrowers a way of understanding themselves in global trade and a code by which to perform. In doing so, it also invokes a set of governmental practices from the pricing of grapes, to quality clauses in supply contracts, and a set of accepted practices mediated by notions of shared fate.

Domestically, this romantic, quality centred identity gives wine and winemakers a disproportionate level of interest in the popular press relative to their national economic contribution, whilst internationally it generates interest in New Zealand disproportionate to its global size and significance. The identity dovetails with images of environmental purity, artisanality, adaptability, competitiveness, creativity, and sophistication. It stands as an almost unique example of how the various and contradictory images of New Zealand promoted by national economic development agencies for consumption at home and abroad might be blended. For these reasons, the NZWI is deployed in projects of economic nationalism and is fostered by these agencies for this reason (see Larner *et al.* 2007).

Recent high profile disputes over practices at wine awards, price cutting, selling juice to supermarkets for buyers own brands, and comparative advertising, confirm that this identity is vulnerable, must be continuously re-made, and is far from a unified governmental force. It is riven with power relations and contests along key axes of difference – among regions and between growers and winemakers, large and small, rich and struggling, and established businesses and new entrants, as well as the different views of individuals. Current increasing levels of foreign ownership and capital concentration promise new pressures, as will any pressures on price (Lewis 2004). The NZW, which has been highly active in fostering this identity, must now ensure that it remains meaningful, affective, and effective.

Conclusion

Although highlighting economic relationality and related interdependencies and their governance, commodity chain analyses only partially capture the constitutive potential of industry. They then impose a linear order upon it. In this chapter I have emphasised the political, the cultural, and the spatial as well as the economic in the constitution of industries – the significance of imaginaries, institutions, and actor-interests as well as economic relations. In this view, the 'New Zealand Wine Industry' is co-constituted by, and gives rise and meaning to, multiple governmental practices – from research to various practices of self-regulation, wine-specific law, the work of industry organisations, trust and family relations, and corporate hierarchies. This interpretation of industry extends beyond commodity chains and governance narrowly writ to encompass the dominant discourses (family, quality, understandings of place, and collective fate) that forge a distinctive industry identity. Place is central and impacts through the various dimensions of *terroir* from winemaking knowledges and practices through to quality aspirations and representations of place in marketing material.

This view of industry takes us beyond the more familiar refrain that industry is a social construction to three observations for the conduct of industry studies. First, 'industry' is one of multiple ways of imagining and governing economic relations. One particular contribution of the chapter in this regard is to direct attention to the role of industry organisations in mediating the interrelationships among enterprises and between them, the state and community; integrating and institutionalising social and professional networks; identifying and representing collective interests; fostering cooperative behavior; and cultivating and discursively constituting industry identities.

The NZW deploys powers given to it both by its members and the state to articulate a rich institutional order of cooperative relations, socio-spatial bonds and shared interests. The centrality of its members within the *filière*, its state sanction, and its active leadership, have all positioned NZW towards the centre of the industrial order. It cultivates Brand New Zealand wine as a central industry identity that emphasises the collective fates of its members and their reliance on sustaining images of place and quality. Its guardianship over the brand links its members to wider projects of economic nationalism.

Secondly, the chapter has drawn attention to the mobilisation of industry as space of governance in modes of governance at a distance – a more immediate and manageable set of economic relations, struggles and identities than those imagined in the language of class, national development, or bureaucracy. It provides a space in which 'neoliberal statecraft' can be practiced (Peck and Tickell 2002). Four elements in particular highlight the significance of industry as a space of governance – the appeal to industries as policy objects for export promotion, frames for naturalising globalisation, post-Fordist subjects of government, and communities of shared fate. The expectations of Brand New Zealand Wine to sell wine and other products of New Zealand on the one hand and the WWTG to facilitate international trade on the other, both link the active constitution of the NZWI to neoliberal political projects. The NZWI is also used to substitute new, class-neutral, post-Fordist work-based identities for capitalists/bosses and workers, such that the very different interests of workers and wine enterprises are elided. The Wine Act is a clear example of neoliberal inspired remote governance through industry self-regulation, whilst the work of NZW has built the industry as a community of shared fate for self-regulation. NZW has assumed a new prominence as a governmental subject in all of this, and is expected to take on the burdens of collaboration, 'delivering' policy, and exhorting growth in the name of the industry. The NZWI case adds to understandings of how as neoliberalism is 'rolled out' as 'after-neoliberalism' and industrial policy reinvented, industries are being linked as spaces of governance to projects of economic nationalism, self-regulation, and globalisation.

Thirdly, the chapter has implications for the practice and politics of industry studies. Industry is clearly one of the 'qualitatively new political economic spaces' that MacLeod (2001) and others argue are emerging from contemporary reterritorialisations of governance, and which need to be studied in such terms. It forms an analytical space in which we might conduct a post-structural political-economy and emphasise relationality, situatedness, and constitutiveness and generate a more fluid conceptualisation of spatiality than scale. The chapter suggests that industries are key spaces for framing not just economic exchange but also socio-economic identities, and political representations and contests. A 'spaces of governance' approach to industry studies highlights broader notions of governance and interdependency, a more political economic order, and the constitutive nature of relational constructs. It should assist researchers to incorporate the cultural and the political in ways that do not over-privilege the economic.

As a final comment, there are two important qualifications to make. First, the aim of the chapter is not to announce or celebrate falsely the appearance of a democratic capitalist paradise. The inner workings of the NZW, for example, are deeply political and riven by contest, whilst 'the industry' is fissured by multiple exclusive practices, especially around labour, new entrants, the landless, and increasingly size (Beer and Lewis 2006). Rather, the aim is to reveal how and why a discourse of shared interest is constructed and sustained, and what this means for the use of the term industry, politically and analytically. In certain conditions (sustained growth, very deep-seated interdependencies, effective governance structures and agency, and links to dominant political projects), such an identity can be sustained as meaningful and formative. That material contests exist is precisely the point. And second, whilst the notion of the NZWI appears naturalised by national borders, it is now clearly and actively carved out of a wider set of international flows of capital, grapes and wine, parts of firms and functions performed and particular employees. This national constitution is sustained by the territorialisation of wine – by regulatory traditions and practices and discourses of *terroir* linking wine production practices and quality associations to place. In the New Zealand case this territorialisation is written into national institutions and discourses, which are now being picked up on by New Zealand's dominant economic nationalism. All this is expressed neatly in the work of NZW and in the notion of BrandNZ wine.

Acknowledgements

This research is supported by a New Zealand Science and Technology Post-doctoral Fellowship and draws on the work of the Wine Industry Research Institute at The University of Auckland.

Reference

Anderson, K. (2004) (eds), *The World's Wine Markets: Globalization at Work*, Edward Elgar, Cheltenham.

Barker, J. (2004), Different Worlds: Law and the Changing Geographies of Wine in France and New Zealand', Unpublished PhD thesis, The University of Auckland, Auckland.

Barker, J., Lewis, N., and Moran, W. (2001), 'Reregulation and the development of the New Zealand wine industry', *Journal of Wine Industry Research*, vol. 12, no. 3, pp. 199–222.

Banks, G., Kelly, S., Lewis, N. and Sharpe, S. (2007), 'Place "from one glance": the use of place in the marketing of New Zealand and Australian wines', *Australian Geographer*, vol. 38, no. 1, pp. 15–35.

Beer, C. and Lewis, N. (2006), 'Labouring in the vineyards of Marlborough: experiences, meanings, and policies', *Journal of Wine Research*, vol. 17, pp. 95–106.

Beverland, M. (2005), 'Crafting brand authenticity: the case of luxury wines', *Journal of Management Studies*, vol. 42, pp. 1003–29.

Coleman, W. (1997), 'Associational governance in a globalizing era: weathering the storm', in Hollingsworth, J. and Boyer, R. (eds), *Contemporary Capitalism: The Embeddedness of Institutions*, Cambridge University Press, Cambridge, pp. 127–53.

Coleman, W. (2002), 'Globalisation and the future of associational governance', in Greenwood, J. (eds), *The Effectiveness of EU Business Associations*, Palgrave, Basingstoke, pp. 209–22.

Friedman, H. (1978), 'World markets, state, and family farm: social bases of household production in the era of wage labour', *Comparative Studies in Society and History*, vol. 20, pp. 545–86.

Gade, D. (2004), 'Tradition, Territory, and *Terroir* in French Viniculture: Cassis, France, and Appellation Controlée', *Annals of the Association of American Geographers*, vol. 94, no. 4, pp. 848–67.

Gibson-Graham, J.K. (2006), *A Post-capitalist Politics*, University of Minnesota Press, Minneapolis.

Herrigel, G. (1994), 'Industry as a form of order: a comparison of the historical development of the machine tool industries in the United States and Germany', in Hollingsworth, J., Schmitter, P. and Streeck, W. (eds), *Governing Capitalist Economies: Performance and Control of Economic Sectors*, Oxford University Press, Oxford.

Hollingsworth, J. and Boyer, R. (eds) (1997), *Contemporary Capitalism: The Embeddedness of Institutions*, Cambridge University Press, Cambridge.

Jessop, B. (1997), 'Capitalism and its future: remarks on regulation, government, and governance', *Review of International Political Economy*, vol. 4, no. 3, pp. 561–81.

Larner, W., Le Heron, R. and Lewis, N. (2007), 'The spaces of "after neoliberalism": co-constituting the New Zealand designer fashion industry', in Keil, R. and Mahon, R. (eds), *Leviathan Undone? Towards a Political Economy of Scale*, University of British Columbia Press, Vancouver.

Le Heron, R. (2006), 'Towards governing spaces sustainably – reflections in the context of Auckland, New Zealand', *Geoforum*, vol. 37, pp. 441–6.

Leyshon A., Lee R. and Williams C. (2003) (eds), *Alternative Economic Spaces*, Sage, London.

Lewis, N. (2003), 'Associational governance and industry making: New Zealand Winegrowers', in Gao, J., Le Heron, R. and Logie, J. (eds), *Proceedings of the New Zealand Geographical Society Annual Conference*, Auckland, July 2003.

Lewis, N. (2004), The New Zealand Wine Industry: Issues and Status at Vintage 2004, WIRI-NZTE Wine Industry Research Project, commissioned by New Zealand Trade and Enterprise, Auckland Uniservices Ltd, Auckland, pp. 130.

Lewis, N. and Kelly, S. (2006), 'Part of the Family? Synergies and sepia in the selling of New Zealand wine', paper presented to the AAG Annual Meeting, Chicago, March.

Lewis, N., Moran, W., Cornet, P. and Barker, J. (2002), 'Territoriality, enterprise and réglementation in industry governance' *Progress in Human Geography*, vol. 26, pp. 433–62.

Lewis, N. (2005), 'Industry as space of government: neoliberal governance and rescaling in New Zealand', *Occasional Publication*, vol. 48, School of Geography and Environmental Science, The University of Auckland, Auckland.

Macleod, G. (2001), 'New regionalism reconsidered: globalisation and the remaking of political economic space', *International Journal of Urban and Regional Research*, vol. 25, pp. 804–29.

Mitchell, T. (2002), *Rule of Experts: Egypt, Techno-Politics, Modernity*, University of California Press, Berkeley.

Moran, W. (1993a), 'Rural space as intellectual property', *Political Geography*, vol. 12, pp. 263–77.

Moran, W. (1993b), 'The wine appellation as territory in France and California', *Annals of the Association of American Geographers*, vol. 83 no.4, pp. 694–717.

Moran, W. (2001), '*Terroir*: the human factor', *The Australian and New Zealand Wine Industry Journal*, vol. 16, no. 2, pp. 32–51.

Moran, W. (2000), 'Culture et nature dans la geographie de l'industrie vinicole Neo-Zelandaise', *Annales de Geographie*, pp. 525–51.

Moran, W. (2006), 'Crafting *terroir*: people in cool climates, soils and markets', International Cool Climate Symposium, Christchurch, 6–10 February 2006.

Moran, W., Blunden, G. and Greenwood, J. (1993), 'The role of family farming in agrarian change', *Progress in Human Geography*, vol. 17, pp. 22–42.

Morgan, K., Marsden, T. and Murdoch, J. (2006), *Worlds of Food: Place, Power and Provenance in the Food Chain*, Oxford University Press, Oxford.

Peck, J. (2002), 'Political economies of scale: fast policy, interscalar relations, and neo-liberal workfare', *Economic Geography*, vol. 78, no. 3, pp. 331–60.

Peck, J. (2004), 'Geography and public policy: constructions of human geography', *Progress in Human Geography*, vol. 28, pp. 392–405.

Peck, J. and Tickell, A. (2002), 'Neoliberalising space', *Antipode*, vol. 34, no. 3, pp. 381–404.

Perry, M. (2001), *Shared Trust in New Zealand*, Institute of Policy Studies, Wellington.

Sheppard, E. and McMaster, R. (eds) (2004), *Scale and Geographic Inquiry: Nature, Society and Method*, Blackwell, Oxford.

Storper, M. (1997), *The Regional World: Territorial Development in a Global Economy*, The Guilford Press, New York.

Streeck, W. and Schmitter, P. (1985), *Private Interest Government: Beyond Market and State*, Sage, London.

Vaudour, E. (2002), 'The quality of grapes and wine in relation to Geography: notions of *Terroir* at various scales', *Journal of Wine Industry Research*, vol. 13, no. 2, pp. 117–41.

Yeung, H. (2005), 'Rethinking relational economic geography', *Transactions of the Institute of British Geographers*, vol. 30, pp. 37–51.

Chapter 10

Placing Local Food in a Cross-Border Setting

Brian Ilbery and Damian Maye

Changing Geographies of Food Supply

The commodity chain concept has been applied to a wide range of food and non-food products. Reflecting increasing consumer demands for high quality and year-round food, corporate retailers are sourcing products from different parts of the world (Barrett *et al.* 2004). The global commodity chain (GCC), which has its roots in world systems theory, has become a key theoretical device for explaining the production of food in 'peripheral' regions of the world economy for retail and consumption in the 'core' (Hughes and Riemer 2004; Jackson *et al.* 2006). This approach has been complemented and extended by the more explicitly political-economic ideas of the global value chain (GVC), with a focus on governance forms and the nature of relations between key actors in a global context (Bourlakis and Weightman 2004; Ponte and Gibbon 2005). However, both approaches tend to 'emphasise large-scale food systems at the expense of the cultural richness of localities' (Hughes and Reimer 2004, 2–3). As Murdoch (1997) suggests food chains or, as he prefers, food networks are always localised, working in real places and at specific times and can only be considered as globalised in terms of their physical extension across space.

Thus globalisation tendencies are mediated by regional and local relationships (Le Heron 1993) and it is global-local interactions that lead to uneven development. Of particular interest here is the response of local actors and institutions to globalising forces, especially given the 'new notions of quality that reflect contemporary concerns with personal health and environmental sustainability' (Goodman and Watts 1997, preface) and the increased emphasis on local food systems as positive mechanisms for regional development. The chapter thus concerns itself with the quality-based, local food sector and especially the relationship this sector has with regional institutions. This is not to say that value chain researchers would deny the importance of local food provision. However, value chain studies understandably concentrate on the global scale because this is where the majority of commodities are channelled, with most arriving on supermarket shelves. Local food has received considerable public support in recent years as a tool to both 'reconnect' producers and consumers and rejuvenate rural/regional economies (Morris and Buller 2003; Watts *et al.* 2005; Ilbery *et al.* 2006). Given this interest, in parallel with a renewed interest in the region more broadly (Marsden and Sonnino 2006), it is surprising that

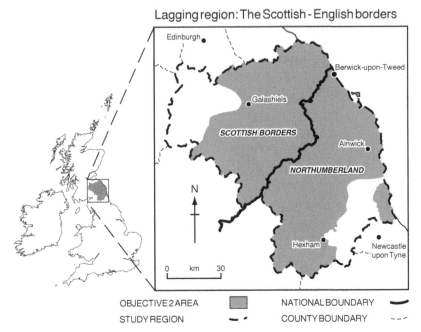

Figure 10.1 of the Scottish-English borders study region text included in image: Lagging region: The Scottish-English borders; Edinburgh; Berwick-upon-Tweed; Galashiels; SCOTTISH BORDERS; Alnwick; NORTHUMBERLAND; N; Hexham; Newcastle upon Tyne; 0 km 30; OBJECTIVE 2 AREA; STUDY REGION; NATIONAL BOUNDARY; COUNTY BOUNDARY; ©MAPS IN MINUTES™ 2003. HMSO Ordnance Survey Permit No. GB399221 ©Crown Copyright, Ordnance Survey & Ordnance Survey Northern Ireland 2003 Permit No. NI 1675 & ©Government of Ireland, Ordnance Survey Ireland.

Figure 10.1   The Scottish-English borders study region

much of the regionalisation debate does not explicitly consider local foods as part of regional development strategies.

This chapter, through a survey of local food chains and related institutional support structures, suggests that such a focus can usefully inform wider debates about regionalisation. More specifically, the chapter focuses on the development of specialist food supply chains in the Scottish-English borders region, a study area which covers two counties: The Borders in Scotland and Northumberland in England (see Figure 10.1). It builds upon previous work exploring supply chain links from producer, intermediary, retailer and institutional perspectives (Ilbery and Maye 2005a; 2005b; 2006; Maye and Ilbery 2006). Using individual supply chains and a 'whole chain' approach, which identifies the geography of supply links and the nature of relationships between nodes in the chain, the key objective is to trace the links between individual producer-led food chains and institutional actors. The chapter also presents an ideal opportunity to explore the potential influence that a national border has on food governance. After providing a brief review of literature on institutional governance and the local food economy, the chapter outlines the 'whole chain' methodological approach and discusses preliminary results from a series of institutional surveys in the region and case study food chains.

Regionalism, Local Foods and Supply Chain Governance

This section provides a brief insight into the two, yet unlinked, literatures on the 'new regionalism' and local foods. There can be little doubt that the number and influence of institutions of regional governance in the UK, and especially England, have increased since the late 1980s (see, for e.g. Jones and MacLeod 2004; Ward and Lowe 2004; Pearce *et al.* 2005). Burch and Gomez (2002) argue that this process has been driven by three factors: first, the need to create regional governance institutions in order to gain access to, and administer, EU structural funds; secondly, the growth of 'new regionalism' among groups of elite regional 'actors'; and thirdly, the creation, in 1994, of government offices in the Regions (GOs). As the powers of GOs increased, two new regional institutions joined them: Regional Development Agencies (RDAs) and Regional Chambers (sometimes called Assemblies). While the former aims to 'improve regional economic performance' (Pearce *et al.* 2005, 198), administer EU structural funds and implement the government's cluster policy, the latter's function is to monitor the activities of the RDAs.

However, there is debate about the extent to which these developments represent a *real* devolution of power to the English regions. For example, evidence suggests that the central government retains considerable influence over regional economic policy and planning (Ward and Lowe 2004). Indeed, Ward and Lowe (2004) identified a paradox in the implementation of the Rural Development Regulation (RDR) in England. While it has strengthened a process of 'Europeanisation' in rural policy development, 'it currently risks being at the cost of distorting sub-national priorities to spread rural development support beyond the farm gate' (Ward and Lowe 2004, 121). The way regions are governed, therefore, may have wide ranging implications for food supply, especially in terms of marketing and organisational support.

The resurgence of interest in foods of local and regional provenance gathered pace in the UK after the publication of the 'Curry Report' (Policy Commission 2002). Responding to the foot and mouth crisis of 2001, this report advocated a need to 'reconnect' farming with the rest of the food chain, including the final consumer. Crucially, Curry recommended that RDAs incorporate regional foods into their economic strategies and put in place the necessary support to enable local food initiatives to nurture and develop. The Department of Environment, Food and Rural Affairs' (Defra) *Rural Strategy* also embraced the principles of de-centralisation for the implementation of economic and social policy for rural areas, including efforts to strengthen working arrangements between regional bodies (Defra 2004). Thus the link between regional governance and local foods is becoming increasingly more explicit.

Policy initiatives to promote foods with territorial associations became a key focus of interest during the 1990s, including the PDO and PGI quality status awarded to dedicated regional foods in the EU (see Parrott *et al.* 2002). Attempts to link 'product and place' in this way became known as the 'turn to quality'. However, other authors contend that the 'localness' of the product is more important than a turn to quality based on, for example, organic or ecological principles (see Winter 2003). Also crucial to this debate is the distinction drawn between 'local' and 'locality' foods; whereas the former refer to products produced and consumed within a certain distance (e.g. 30 miles), the latter relate to products from further afield, but with an identifiable geographical provenance e.g. the sale of Newcastle Brown ale in a supermarket in Coventry or Auckland.

There is little doubt that policy makers and regional agencies are beginning to investigate the economic development potential of the local food sector (see Marsden and Sonnino 2006). Indeed, DuPuis and Goodman (2005) claim that localism in Europe has emerged in the context of new forms of devolved rural governance alongside the slow process of wider European reform of the Common Agricultural Policy (CAP).

Adopting a 'Whole Chain' Methodology

This chapter adopts a 'whole chain' approach to the study of food supply chains. Essentially qualitative in nature, a whole chain approach seeks to identify links upstream and downstream from the producer and to explore the geography and nature of business relations between nodes in the chain. More specifically, the approach aims to:

- Investigate how food supply chains are constructed by specialist producers;
- Trace links between producers and other actors in the supply chain;
- Understand how different actors in the supply chain relate to one another, and;
- Unpack how social and economic relations co-relate in the context of a region's local (quality) food economy.

The main methodological device in the whole chain approach is the drawing of *supply chain diagrams*. As part of in-depth interviews with 43 specialist food producing businesses in the Scottish-English borders region, each producer was asked to draw a supply chain diagram for their business, including both upstream suppliers and downstream links. Once these links were drawn, interviewees were asked to openly talk about the nature of those links, when and how they were established. The core aim, therefore, was to map the supply chain geography of individual food businesses. This chapter focuses on case studies of two livestock producers; these involved repeat visits and discussions over a two year period, thus allowing a more detailed account of the changing nature of individual business supply chains. Each case study examined the supply links by talking to different people associated with the producer/product at different points along the food chain, as well as exploring any significant links with key institutions that have supported the business in some way.

However, the chapter first introduces results from an institutional survey conducted with 25 regional and national institutions (Table 10.1a and b). Selected according to institutional type (e.g. public, charity and private), remit (e.g. economic, rural/ regional development), links with the case study region and relevance to food SMEs, face-to-face interviews were conducted with key individuals within the institutions. A key component of the interview involved asking respondents to draw their own institutional map. These diagrams were analysed to identify institutional links to the food chain and the nature of food governance within the study region.

Table 10.1a Institutional profile – England/The North East/Northumberland

Institution	Start	Principal Activity	Geographic Scope
f3 – Foundation for Local Food Initiatives	1999	Promote/support the growth of healthy local food economies. Consultancy, training and research	N
Sustain: the alliance for better farming and food	1997	Support/encourage sustainable food and agricultural policies and practices. Core policy work and a series of reports/projects	N/I
Local Food Works (Soil Association)	2002	Foster sustainable systems through local food networks. Information and network development services	N
Institute of Grocery Distribution (IGD)	1909	Research and education services	N
Meat and Livestock Commission (MLC) (catering)	1967	Food service: represent producers and satisfy customer demands. Public sector: promote red meat	N
English Beef and Lamb Executive (EBLEX)	2003	Advocacy for cattle and sheep producers in England	N
Government Office for the North East (GONE)	1994	Co-ordinates delivery of government policy in the region	R
One NorthEast (ONE)	1999	Responsible for developing/implementing a regional economic strategy	R
Department of Environment, Food and Rural Affairs (DEFRA)	2001	Rural Development Service delivers the England Rural Development Plan (ERDP)	R/N
Northumbria Larder (NL)	2001	Regional food group for speciality producers in the region	R
The Countryside Agency (CA)	1999	Statutory advisor on countryside recreation and conservation	R/N
North East Land Links (CA sponsored)	2001	Develops countryside and community linkages	R
Business Link Northumberland (BLN)	1992	Promotes economic and business development	L
Northumberland Farmers' Market Association (NFMA)	2000	Helps promote, publicise and make permanent farmers' markets in Northumberland	L
Northumberland County Council (catering)	-	Oversees catering services/contracts for schools and other public bodies	L
Hadrian's Wall Tourism Partnership (HWTP)	1995	Promotes sustainable tourism and economic development along the wall	Sub-R

L = Local R = Regional N = National

Governing Food Supply Chains in the Scottish-English Borders

This section outlines the key mechanisms that influence the governance of the region's local/specialist food sector, contrasting the situation in The North East/ Northumberland with that in Scotland/The Borders. Surveyed institutions operate on three different scales: first, absorbing international and national *policy* agendas for regional/rural governance; second, transforming those policies into regional/local *strategies*; and third, establishing relevant *projects* at the regional/local level. It is recognised that scaling food chain governance in this way presents a stylised synopsis, but it helps to make sense of a confusing and complex set of arrangements.

In the North East/Northumberland, there is quite a stratified division of labour in terms of national policy, regional strategy and local projects. The policy part of the governance framework lies 'outside' the study region and the current overarching national public policy driver in terms of food is the UK Government's *Sustainable Farming and Food Strategy*. The strategic scale involves the incorporation of national policies into regional strategies. Four institutions are significant to food supply chains in the North East region. Of these, GONE and ONE play the most important roles (Table 1a). In response to Curry (2002), ONE has been tasked (along with GONE) to devise a regional/local food strategy. The contents of this strategy (or delivery plan) include input from other partners involved in the regional steering group (e.g. Defra, The Countryside Agency). It is significant that each agency in the region has its own regional/local food agenda.

The third scale represents the implementation of regional strategy into 'tangible' local food projects. Projects can be divided into regional (North East) and local (Northumberland) food initiatives. At the regional scale, two projects are significant: first, Northumbria Larder (NL), the region's speciality food and drink group; and secondly, North East Land Links, which is dedicated to two main strands of work – public procurement and community food (e.g. food co-ops, community cafes, food delivery schemes). In Northumberland, Business Link Northumberland (BLN) helps local food businesses to prepare grant applications and attract ERDP funding from Defra. BLN's 'Made in Northumberland' scheme promotes and markets speciality food products, and organises local trade fairs. Other local food projects are also on-going in the region, including, for example, Hadrian's Wall Tourism Partnership.

There is a less stratified division of labour in Scotland/The Borders. In terms of policy, the influences and directives are broadly similar (Table 10.1b). What is also significant is the continued link with the Cabinet Office in London which, although not determining policy and strategy directly, still retains notable influence. Policy, strategy and projects are all in evidence, with Scotland acting as the regional institutional layer. *A Smart and Successful Scotland* (2001) is the current overarching economic development framework and quality food production is a key economic objective, in particular focusing on developing 'value added strategies'. In terms of food-related projects, Scottish Enterprise Borders is the main institution, with its *'New Ways'* economic development strategy (see Table 10.1b). It is also the key player in terms of business support for food SMEs. Influenced by national policy, emphasis is placed on those businesses willing and able to sell 'out with' The

Borders. The other key institution – the Scottish Borders Council – provides support and marketing, including the production of a *Flavours of The Borders* booklet.

Table 10.1b Institutional profile – Scotland/The Borders

Institution	Start	Principal Activity	Geographic Scope
Scottish Executive Environment and Rural Affairs Department. (SEERAD)	1999	Rural development, financial support for local/quality food sector	N
Seafish (Scotland)	1981	Develop the fishing industry and promote the consumption of fish	N
Farm Retail Association (Scotland)	1994	Promote local food and rural industry	N
Quality Meat Scotland (QMS)	2000	Industry development: promote quality assurance through the chain	N
Scottish Enterprise Borders (SEB)	1994	Economic development – increase GDP via support stream	C
Scottish Borders Council (SBC)	1996	Planning and Development Department: stimulate economic activity Environmental Health Department: enforce food, health and safety legislations	C
Borders College (Chef School)	1971 (2002)	Chef School: training and development; support Border food	C
Borders Farmers' Market Committee	2000	Develop FMs in The Borders, promote local producers	C
TFC Food Services	1992	Food/cookery consultancy: training, promotion and product development	C

C = County N = National

The level of local food activity and related support, therefore, is less in The Borders than in Northumberland. Significantly, on either side of the national border one finds two fairly distinct systems of governance for their respective local economies. While both counties respond to the *Sustainable Food and Farming Strategy*, a range of institutional interests and competing agendas appear to exaggerate the border effect.

Following Livestock Supply Chains in the Scottish-English Borders

The chapter now provides short, descriptive cameos of two case study livestock businesses. Each case study introduces key intermediaries and retailers that are connected to the business and then provides a detailed appreciation of food chain links, whether they have changed and why.

Case Study 1: Organic Hill Meat

This organic farm and on-farm butchery is located in the Northumberland uplands. Established in 1998, the business employs four full-time staff and produces and retails a range of organically reared meat products, using local branding and traceability (see Figure 10.2). Heather and clover-fed lamb is the main product and each cut of meat has its own individual code to guarantee traceability through the supply chain. As the producer explained: 'The whole point about a differentiated product is that it is 100 percent squeaky clean ... we can trace every joint of meat back to the particular animal and a little bit further than that, certainly with the sheep products'.

Constructed to add and retain value from primary production, the business also sources cattle, sheep, pigs and poultry from other organic farms in the region. Where possible, upstream supplies are sourced from within the region; the one main exception is organic feed which comes from across the border, in Kelso. All livestock are slaughtered at the (organically accredited) abattoir in Whitley Bay and then delivered to the on-farm butchery to be processed and packaged for retail. The abattoir owner revealed how this service for specialist producers in the region could not exist without the buttress of established intermediate channels, including supplying wholesale products to catering and butchering customers in the region. The meat products are sold through various short food chains e.g. direct sales, local/regional specialist food shops and caterers (Figure 10.2). Most of the organic meat products are thus sold to retail customers in the region, with only five percent going out of the region.

A business website provides excellent information about the business, the farmer, the products and the butchery. Uptake of on-line meat sales has been slow and the business has also reduced sales output at farmers' markets and stopped supplying (specialist) butchers in the region; the preference now is for more 'stable' outlets (independent retail and catering). The most recent addition, two restaurants in Newcastle and Ponteland, both owned by a leading chef in the region, reflects this growing demand. The chef revealed that the link with the organic business is based firmly on the quality of the product. Two retailers that source organic meat from the farm were also interviewed. The village store in Belsay, less than five miles from the farm, is now one of the farm's best customers and the shopkeeper noted that selling local products, particularly local meat, has helped to save the shop from closure. The small 'ethical supermarket' in Newcastle is owned by a consumer co-operative (17,000 members) and the manager explained that they endeavour to source local (organic) product, but usually cannot because of a lack of suitable producers.

The producer is well networked with relevant institutions, food groups, and the regional and national media for promotion and marketing. Permission has recently been granted to build a new licensed cutting plant, eventually providing a meat cutting service for other farms in the region. Significantly, 10 percent of business profits are re-invested back into the land through a series of conservation projects. Nevertheless, much of the generated value added is lost to high distribution costs,

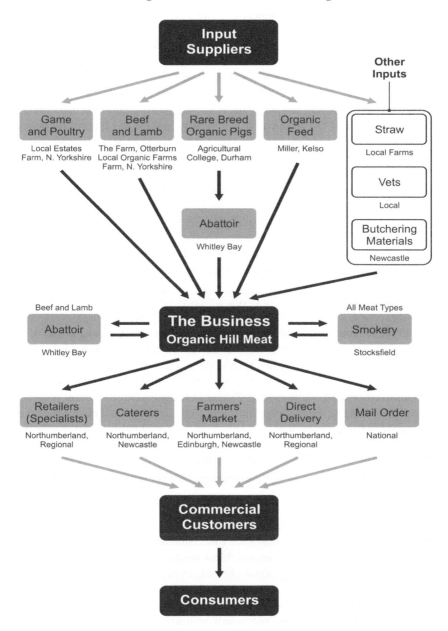

Figure 10.2   Business supply chain diagram for organic hill meat

reflecting the often large distances between the producer and retailers/consumers. Other threats include the closure of the only local abattoir and the relatively low local demand for organic meat products.

Case Study 2: Specialist Cheese and Ice Cream

Located in Glendale Valley, north Northumberland, this dairy farm has been run by the family for over 50 years. The business employs 16 staff (10 full-time and 6 part-time) and produces a range of unpasteurised farmhouse cheeses (since 1990) and specialist ice creams (since 2000). Food processing is a diversification strategy, adding value and enabling family members to remain working on the farm, helped by a keen interest in the product: 'It was mainly interest to be perfectly honest, interest in a quality product [cheese] and development of another enterprise to provide income and employment here on the farm'.

On-farm production comprises two separate, but linked, product supply chains (Figure 10.3a and b). Both products use milk from the farm's 180 dairy cows, but other inputs come from a range of local, regional, national and even international suppliers. Thus national borders are no obstacle as some inputs come from Scotland, France and the Netherlands. For example: 'There are very few people in Britain still doing it. For the cheeses we are making, we have to go to where the technology is ... Some of the companies are so big that you cannot buy orders direct from them, you have to go through someone else'.

Once processed, the majority of cheese products are sold to regional and national (specialist) cheese wholesalers (Figure 10.3a) e.g. Ian Mellis, Edinburgh; Neal's Yard Dairy, London/ New York. Two wholesale businesses were interviewed as part of the study. A cheese wholesaler/delicatessen owner based in Northumberland described how 'local cheeses' are sourced direct from two farms in Northumberland (including the case study farm), as well as from other farms in Yorkshire. The business also uses major national wholesalers (e.g. Cheese Seller, London) to source 'non-local cheeses', and obtains French cheeses direct from an agent in France. However, local cheeses offered the business 'an important point of difference' when selling to hotels and caterers in the region. Another cheese wholesale business owner, based in Edinburgh, explained how he sourced specialist cheeses direct from local producers all around the British Isles, as well as mainland Europe. For him, the quality of the cheese is the overriding factor.

Approximately two-thirds of the cheese products from the case study business are sold to customers in the region, with 30 percent sold nationally and 5 percent internationally. In contrast, the geographic dispersion for ice cream is more localised, with 100 percent sold within the region. Most ice cream is sold to independent and specialist retailers (some also source cheese), as well as to five regional ASDA stores as part of the latter's 'local choice' food initiative (Figure 10.3b). An ASDA store manager in Gateshead and the manager of the regional sourcing team in Leeds confirmed this link with the farm. The store manager praised the producer's excellent product and marketing ability. As well as ASDA, a number of other surveyed retail customers in the North East sell the farm's dairy products, particularly ice cream. This includes the Fenwick's department store in Newcastle. Unlike its supermarket counterparts, Fenwick's has been promoting and sourcing local/regional products for a number of years; the food buyer explained how such products provided '... good points of difference from multiples'. However, as he also testified: 'All the multiples now want to go heavily into regional products'.

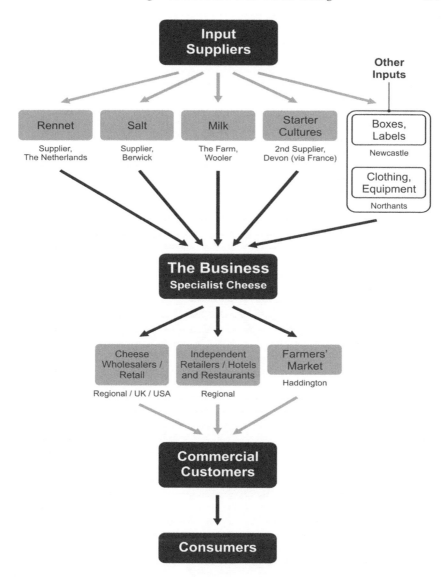

Figure 10.3a   Business supply chain diagram for specialist cheese

The farm has significantly reduced attendance at farmers' markets as a result of poor sales, logistical problems and cost constraints. High distribution costs are a weakness of this business, but the major threat relates to any bovine TB scare in the milk herd. To reduce this risk from raw milk cheese production, one possible option is to produce a pasteurised cheese.

To summarise this section, both case study businesses have developed identifiable regional products, with a clear focus on high quality production standards. The farms

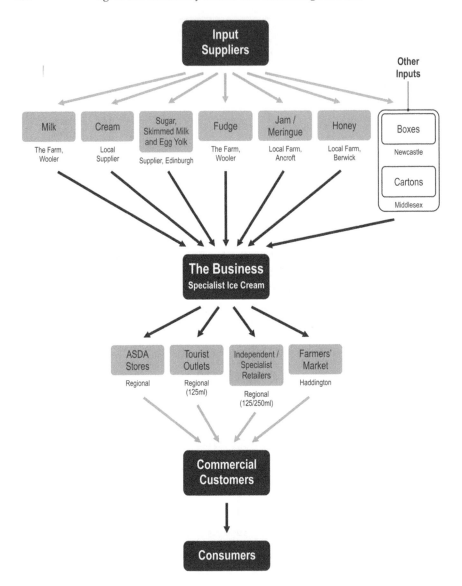

Figure 10.3b   Business supply chain diagram for specialist ice cream

also co-operate with other like-minded producers and have established a good local/ regional supply base. For the organic meat business, the supply chain is largely contained within the region, accessing its inputs and marketing its outputs in the local area. In contrast, the dairy business, especially its cheese production, is forced to look outside the county for both input supplies and to sell the product. For the dairy business, as with many other case studies in the survey, the Scottish-English border is almost irrelevant, except perhaps in terms of accessing institutional support. Both case studies also reveal, through following their supply links, complex intermediate and retail food geographies. Once again, the national border is not significant and emphasises the elastic nature of local foods as they move through the supply chain to be processed, marketed and/or sold by intermediaries and retailers.

Conclusions

This chapter began by suggesting that, despite continued interest in global commodity and value chains, globalisation tendencies are always mediated by regional and local relationships. Indeed, renewed interest in local and regional foods in the UK and many other parts of the developed world has coincided with a growth in new instruments of regional governance. As a consequence, a number of regional agencies are now investigating the economic development potential of the local food sector. Spurred on by academic interest in both commodity supply chains and local food supply systems (see Jackson *et al.* 2006 and Maye and Ilbery 2006 for recent reviews), the chapter has examined some of the links between individual (specialist) food chains and institutional actors in the Scottish-English borders region. In particular, it has used a 'whole chain' methodology to show how specialist livestock businesses construct and operate highly individual and customised supply chains.

Empirically, the chapter began with a survey of institutions in both border counties. Various issues flowed from this analysis, with similarities in terms of the initiatives in place (or at least the philosophy behind them), as well as the duplication of effort and an over-abundance of regional economic strategies mixed with competing institutional views. However, more significant is the difference in the way the two counties' food governance arrangements are organised, with a fairly stratified layering according to the '3-scale' policy, strategy, project framework in the North East/Northumberland, compared to a less stratified framework in Scotland/ The Borders. In the latter especially, there is also a clearer centralisation of policy at the local level.

Various issues also emerged from the two livestock case studies in terms of the cross-border situation and the value of adopting a 'whole chain' approach. First, it is clear that local supply chains do not start at the point of production (see also Ilbery and Maye 2005 a and b). In the case of the dairy business, the whole chain perspective takes the supply chain outside the study region and into mainland Europe. Second, local businesses are intricately 'bound up' with various downstream actors, including abattoirs, carriers, wholesalers and retailers. In this situation, localness, in terms of where a product is produced, is even more contested. Retail surveys, for instance, reveal a market typified by intra-sector competitive dynamics when it comes to dealing with local products, especially as supermarket interest increases. Broadly, the role of the national border seems less significant further down the supply chain. One or two producers did report a border effect, but the territorial identity of certain study products is ephemeral, with the same product marketed in specific ways to identify and cater for certain retail markets. Third, both case study businesses have changed their supply chain arrangements in the two year survey period, especially in terms of reducing attendance at farmers' markets in pursuit of more 'stable' retail markets. This shows the value of a longitudinal methodology, with repeat visits; this is an important element of a 'whole chain' approach.

Overall, therefore, the findings appear to show a mismatch between institutional support (where there is a notable border effect) and the nature of producer supply chains, which tend to be organised more on the basis of economic imperative and informal supply chain arrangements than according to the existence of a national border. This is perhaps not surprising. In both cases, specialist producers do make a concerted effort to sell their products locally, but this is rarely sufficient on its own. Finally, the process of drawing and following different livestock supply chains (as well as compiling a broad outline of the surrounding institutional landscape) is valuable in terms of helping to interrogate notions of 'localness', as well as potentially informing future cross-border studies of food supply.

Acknowledgements

This chapter derives from the EU-funded project: 'Supply chains linking food SMEs in Europe's lagging rural regions (SUPPLIERS, QLK5-CT-2000-00841). Collaborating laboratories are: SAC, Aberdeen, UK (Co-ordinator); Coventry University, UK; University of Wales, Aberystwyth, UK; Teagasc, Dublin, Ireland; ENITA, Clermont-Ferrand, France; University of Patras, Greece; SIRRT, University of Helsinki, Finland; and the Agricultural University of Krakow, Poland. The helpful comments made by two anonymous reviewers on an earlier draft of this chapter are also acknowledged with thanks.

References

Barrett, H., Browne, A. and Ilbery, B. (2004), 'From farm to supermarket; the trade in fresh horticultural produce from sub-Saharan Africa to the United Kingdom', in Hughes, A. and Reimer, S. (eds), *Geographies of Commodity Chains*, Routledge, London, pp. 19–38.
Bourlakis, M. and Weightman, P. (2004), *Food Supply Chain Management*, Blackwell, Oxford.
Burch, M. and Gomez, R. (2002), 'The English regions and the European Union', *Regional Studies*, vol. 36, pp. 767–78.
Curry Report (2002), *Farming and Food: A Sustainable Future*, Policy Commission on the Future of Farming and Food, Cabinet Office, London.
Defra (2004), *Rural Strategy*, Defra, London.
DuPuis, M. and Goodman, D. (2005), 'Should we go "home" to eat?: toward a reflexive politics of localism', *Journal of Rural Studies*, vol. 21, pp. 259–371.
Goodman, D. and Watts, M. (eds) (1997), *Globalising Food: Agrarian Questions and Global Restructuring*, Routledge, London.
Hughes, A. and Reimer, S. (2004), 'Introduction', in Hughes, A. and Reimer, S. (eds), *Geographies of Commodity Chains*, Routledge, London, pp. 1–16.

Ilbery, B. and Maye, D. (2005a), 'Alternative (shorter) food supply chains and specialist livestock products in the Scottish-English borders', *Environment and Planning A*, vol. 37, pp. 823–44.

Ilbery, B. and Maye, D. (2005b), 'Food supply chains and sustainability: evidence from specialist food producers in the Scottish/English borders', *Land Use Policy*, vol. 22, pp. 331–44.

Ilbery, B. and Maye, D. (2006), 'Retailing local food in the Scottish-English borders: a supply chain perspective', *Geoforum*, vol. 37, pp. 352–67.

Ilbery, B., Watts, D., Simpson, S., Gilg, A. and Little, J. (2006), 'Mapping local foods: evidence from two English regions', *British Food Journal*, vol. 108, no. 3, pp. 213–25.

Jackson, P., Ward, N. and Russell, P. (2006), 'Mobilising the commodity chain concept in the politics of food and farming', *Journal of Rural Studies*, vol. 22, pp. 129–41.

Jones, M. and MacLeod, G. (2004), 'Regional spaces, spaces of regionalism: territory, insurgent politics and the English question', *Transactions of the Institute of British Geographers*, vol. 29, pp. 433–52.

Le Heron, R. (1993), *Globalised Agriculture: Political Choice*, Pergamon Press, Oxford.

Marsden, T. and Sonnino, R. (2006), 'Rural development and agri-food governance in Europe: tracing the development of alternatives', in Higgins, V. and Lawrence, G. (eds), *Agricultural Governance: Globalisation and the New Politics of Regulation*, Routledge, London, pp. 50–68.

Maye, D. and Ilbery, B. (2006), 'Regional economies of local food production: tracing food chain links between 'specialist' producers and intermediaries in the Scottish-English borders', *European Urban and Regional Studies*, vol. 13, pp. 337–54.

Morris, C. and Buller, H. (2003), 'The local food sector: a preliminary assessment of its form and impact in Gloucestershire', *British Food Journal*, vol. 105, pp. 559–66.

Murdoch, J. (1997), 'Towards a geography of heterogeneous associations', *Progress in Human Geography*, vol. 21, pp. 321–37.

Parrott, N., Wilson, N. and Murdoch, J. (2002), 'Spatializing quality: regional protection and the alternative geography of food', *European Urban and Regional Studies*, vol. 9, pp. 241–61.

Pearce, G., Ayres, S. and Tricker, M. (2005), 'Decentralisation and devolution to the English regions: assessing the implications for rural policy and delivery', *Journal of Rural Studies*, vol. 21, pp. 197–212.

Ponte, S. and Gibbon, P. (2005), 'Quality standards, conventions and the governance of global value chains', *Economy and Society*, vol. 34, pp. 1–31.

Scottish Executive (2001), *A Smart and Successful Scotland: Ambitions for the Enterprise Networks*, Scottish Executive, Edinburgh.

Ward, N. and Lowe, P. (2004), 'Europeanizing rural development? Implementing the CAP's second pillar in England', *International Planning Studies*, vol. 9, no. 2–3, pp. 23–32.

Watts, D.C.H., Ilbery, B. and Maye, D. (2005), 'Making re-connections in agro-food geography: alternative systems of food provision', *Progress in Human Geography*, vol. 29, no. 1, pp. 22–40.

Winter, M. (2003), 'The policy impact of the foot and mouth epidemic', *The Political Quarterly*, vol. 74, pp. 47–56.

Chapter 11

Agrarian Clusters and Chains in Rural Areas of Germany and Poland

Elmar Kulke

Introduction

At present, the analysis of clusters and commodity chains represents a central research area in the field of economic geography. In a globalising world, classical spatial structures of the economy – which were previously characterised by the strong influence of the nation-state – are changing. Supra-regional and international relations are becoming more and more important; these linkages based on material and immaterial flows can be analysed in their spatial dimension using the commodity chain approach. And within a globalised system, local concentrations of companies with intensive networks are also gaining competitive advantages; these concentrations can be described and analysed by network and cluster models. For the most part, studies analysing clusters and commodity chains are oriented towards manufacturing industries and service activities. For some time there has been a tendency in economic geography to neglect the agricultural sector and rural areas. This may be based on the fact that, in advanced economies, agriculture is – in terms of employment – of very limited economic importance, and that rural areas are labelled as an economically weak periphery.

This study attempts to transfer the knowledge of clusters and commodity chains found in the analysis of the manufacturing sector to the agricultural sector. It is based on empirical studies performed by a research group, with support from the German Research Foundation, of structural change and transformation in agriculture in Germany and Poland.

The basic hypothesis is that competitive clusters of agriculture and complementary services also emerge in rural areas, and are significant for regional economies. Elements of this cluster formation are not only material linkages but, in addition, immaterial connections based on the flow of information and co-operation in projects. The second hypothesis is that not all agricultural units are integrated into the cluster network to the same extent. The analysis shows that, depending upon certain typical characteristics of the farms, different intensities in their level of integration into the cluster can be observed; it also documents that an intensive integration into the local cluster, in combination with supra-regional linkages, improves a farm's competitive position.

This chapter will first introduce the models of clusters and product chains while describing their most relevant aspects with respect to rural areas. These aspects will then be analysed empirically on the basis of two case studies in Germany and Poland.

The Concept of Clusters and Chains in Economic Geography

The fundamental idea behind cluster approaches is that spatial proximity between companies and supplementary units (e.g. institutions, suppliers, services) not only leads to cost reduction through classical agglomeration advantages (transport and transaction costs) but also opens up possibilities for the development of intensive immaterial linkages (Bathelt and Glückler 2002; Granovetter 1985; Kulke 2004). The exchange of information – especially of non-codified, experience-based tacit knowledge – based on face-to-face contact between actors can lead to a mutual learning process. This exchange is strong if the network is rooted in reciprocity, trust, dialogue and flexibility, and if it is embedded in a common human, cultural, social and political environment.

Lead by this idea, different cluster approaches have been developed based primarily on observations made of manufacturing industries. The 'innovative milieu' and the 'geographical industrialisation' approaches focus on the spatial concentration of modern and innovative production (e.g. Aydalot 1986; Camagni 1991; Storper and Walker 1989); the explanatory value of these approaches for classical agricultural production, however, is very limited. In contrast, the models of industrial districts (based on studies by Marshall 1927) and of regional industrial clusters (based on the concept by Porter 1993) can be applied not only for manufacturing but for other sectors as well.

The transfer of these cluster approaches forms the theoretical foundation of this study. The basic idea is that structures comparable to those in manufacturing industries may exist in the agricultural sector in rural areas as well. Such agrarian clusters consist (Figure 11.1) of farms which are not only supported by networks with one another but also by relations to preliminary units (e.g. suppliers), to a variety of different services (e.g. maintenance, finance, education) and to downstream units (e.g. wholesalers, further processing). Farms which are integrated into this cluster, with material input-output relationships and immaterial information flows, may have a better competitive position than only weakly linked units. The knowledge of the existence of such an agrarian system could be a starting point for a more comprehensive regional economic policy for rural areas, one which is no longer strictly restricted to agriculture.

An analysis of the quality and competitiveness of a cluster must also include an evaluation of the networks which exist outside the region (Figure 11.1). The concepts of commodity and product chains provide a useful support for this analysis (Gereffi 1996). A commodity chain is defined by the linkages between the raw material, the various steps of production, the distribution and the final consumption. The constitutive elements of the chain are the existence of material input-output relations, of information flows and of the occurrence of an imbalance of power between the participating actors. Generally, these chains are differentiated by the distribution of power. Producer-driven chains appear with companies that own a quasi-monopoly

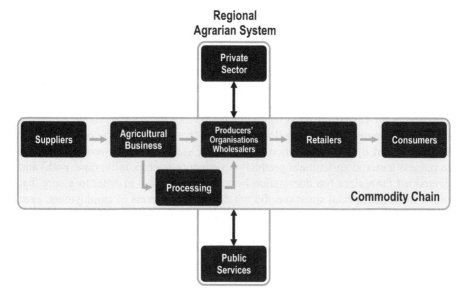

Figure 11.1   Connection between the regional agrarian system and the agrarian commodity chain

Source: Dannenberg and Kulke 2005.

as a result of either modern technology or brand values. Customer-driven chains are often found if large retailers are linked with many producers of similar products; then they are able to implement their price and product ideas due to their purchasing dominance. In agriculture, depending upon the standardisation of the products, a high degree of external dependence is expected. In the best-case scenario, the existence of a cluster improves the products and achieves higher competitiveness through the mutual exchange of information.

From this general discussion, two central questions for the study were derived:

- To what extent are rural areas dominated by agricultural cluster structures based on intensive networks and integration into production chains?
- Which types of agricultural units – depending upon their characteristics and the form or intensity of their integration into the cluster and chain – have a better competitive position?

Based on the cluster and commodity chain theory, this analysis follows an interdependent approach. It is estimated that, especially in agriculture, competitiveness can be achieved if an existing intensively linked local cluster is integrated into supra-regional networks).

Empirical Analysis of Agricultural Clusters and Chains

The following interpretation is based on the empirical results of questionnaire surveys (N = 332 farms) and expert interviews (N = 50; with stakeholders, politicians, institutions) in two case study areas in eastern Germany and Poland; the study was co-ordinated by E. Kulke, and the empirical work and data analysis were carried out by P. Dannenberg (Dannenberg 2006; Dannenberg and Kulke 2005). The selection of the case study areas was based on the idea that they should be situated a relatively large distance from agglomerations, so that rural structures could not have been deformed by suburbanisation tendencies. Case study areas were also chosen to possess a mixed agricultural production system (e.g. crop, milk, rape, pork) and diversified farm sizes (no domination by a few large units) in order to assure that mono-structural special situations did not dominate. In terms of employment, sites were sought which displayed agriculture as being of high importance (6.6 percent of the workforce in the selected areas in Germany, 22.6 percent in Poland).

Because of the development path of the 20th century, differences between the two countries can be observed (Pawlak 2004; Pfeiffer 2004). In Poland there has been a long-time persistence of private structures which originated before the socialist period and have persisted until the market economy of today; therefore agricultural production is dominated by smaller private-owned full-time farms. In East Germany during the socialist period, agriculture was concentrated in a few very large production units (LPGs). These large socialist agricultural production units were subdivided and privatised during the economic transformation process of the 1990s. Today a polarised farm size structure can be observed. While there are small private units which are often run as part-time farms by the former owners of the land, there are also relatively large (between 100 ha and 1000 ha), modern companies which were newly formed during the privatisation process and are run either as partnerships or as corporations.

Linkages and Cluster Formation in Agriculture

The data of the survey show that in both case study areas, most of the farms have strong local (district/county) and regional (state/province) linkages. The largest share of the turnover produced by the farms remains in the region. More than 60 percent of all farms in both areas work together with local suppliers (Table 11.1). Local residents of the area comprise over 50 percent of the direct customers and service providers. Less than 10 percent of the interviewed farms have more than 50 percent of their suppliers, customers or service providers outside of eastern Germany or Poland. In this context it is important to mention that most of the farms do not work together with all types of service providers: Almost 90 percent of the farms work together with insurance companies, while only 36 percent work together with marketing companies (banks 85 percent, handicrafts 77 percent, agricultural consulting 72 percent, legal consulting 53 percent, computing 43 percent). Standardised services are mainly used within the region, while linkages with modern knowledge-intensive services show a higher supra-regional share.

Between the different types of farms and the two regions, different spatial networks can be observed. In general, the small units in Poland show the highest share of local contact with suppliers and buyers; often these businesses are so small that they cannot afford a car or truck and are therefore dependent on local or mobile salesmen and buyers. Smaller units tend to have a strong regional orientation and a limited number of contacts. Younger businesses (less than five years old) have fewer local suppliers and buyers than average, and stronger supra-regional linkages, which is partly due to their more intensive use of new equipment and modern production technology. Large units and the newly formed corporations have more contacts and both strong local and supra-regional linkages.

The survey shows that the majority of the businesses organise the predominant part of their supply, sales and service connections within the region (Table 11.1). More than half of the sales of the polled farms with suppliers, buyers and services take place within the region. To identify a local cluster, it has to be taken into consideration that the survey only documents the direct connections in the commodity chain. This allows no statement about how much of the preliminary product is processed and consumed outside the region. However, the analysis shows that, apart from the agricultural production, the overwhelming majority of the agriculture-oriented services and large parts of the directly connected production stages are also located within the region. Due to these material linkages, the term 'regional agrarian cluster' is justified in both regions.

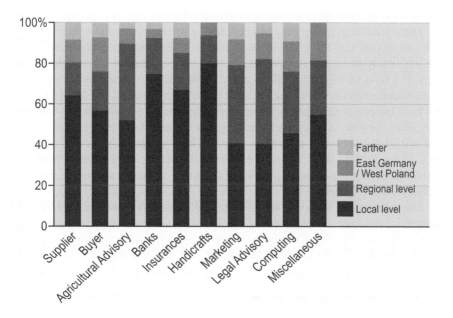

Figure 11.2   Regional distribution of suppliers, customers and service providers for the farms surveyed

Source: Questionnaire surveys (Dannenberg and Kulke 2005).

Up to now the discussion has only analysed material linkages, but for the formation of a competitive and innovative cluster, immaterial linkages are also important. The analysis of these immaterial networks was based on questions in the survey about the extent to which farmers received relevant information through their linkages, and whether they are working together on projects or corporate strategies with other stakeholders. About two-thirds of the businesses surveyed showed various forms of these immaterial linkages, and twenty percent stated that they are using them intensively.

Between the different types of farms, significant differences can be observed. Especially the larger units located in Germany, the privatised GDR LPGs, have more and more intensive linkages with other stakeholders. This can be explained by the experienced and locally embedded management; most of the leading staff have held their jobs for a long time and possess well-established connections with all kinds of local services, institutions and processors. And nearly all managers are integrated into the cattlemen's association, the farmers union or the local fire brigade. This social capital outlasted the transformational period. And finally, it is easier for the larger units to utilise intensive immaterial linkages due to their larger personnel and material resources. Joint projects, the most intensive form of co-operation, are especially undertaken by companies with an annual sales volume of more than €500,000. Depending upon the age of the larger units, differences in their form of co-operation can also be observed; the younger units (which were newly established after the economic transformation) are more interested in consulting and training, and show a higher share of supra-regional connections.

The smaller units in Germany and Poland display less integration with immaterial linkages. In Germany this is due to the high share of part-time farmers who are doing the job simply to be independent and to earn some additional income. In Poland there is a kind of mistrust of all forms of co-operation (Chloupkova *et al.* 2003, 250), which originates from the socialist period. Furthermore, the small units do not have the personnel capacity for co-operation outside of their work in the farm. Very often they buy all their preliminary products (seed, fertiliser, pesticides) from and sell all their products to only one partner; and outside their material linkages, they do not practice other forms of exchange. Therefore, these types of farms are highly dependent – in terms of prices and qualities – upon big agribusinesses, food industries and retailers.

The results show the existence of cluster-like structures based on material and immaterial linkages. But the level of involvement is dependent upon the different types and characteristics of the units. Larger farms are obviously better integrated, whereas small and part-time farms have fewer and weaker linkages.

The Competitive Position of the Farms

In the survey, farms were requested to estimate their competitive position. The evaluation was based on their bargaining position to suppliers and buyers and on their access to relevant business information within the network of immaterial linkages. The competitive position was then correlated with a group of farms which had been formed by a statistical cluster analysis (Table 11.2). Groups 1 and 2 are dominated by the larger units in Germany, former LPGs which were privatised during the period of economic transformation. Group 3 integrates smaller units in Germany and to a lesser extent Poland. Group 4 comprises the newly established larger units in Germany and Poland. Groups 4 and 5 consist mainly of older and smaller units in Poland.

Table 11.1   Cluster analysis based on farm characteristics

Groups (No. of Farms)	1 (40)	2 (13)	3 (75)	4 (26)	5 (137)	6 (24)
Region	Germany	Germany	Mostly Germany	Mixed	Mostly Poland	Poland
Sales Volume	High	High	Low	High	Low	Low
Range	Specialised	Specialised	Mostly specialised	Specialised and mixed	Mixed	Specialised
Legal Form	Mostly business partnerships and capital companies	Mostly business partnerships and capital companies	Individual enterprises	Mostly individual enterprises	Individual enterprises	Individual enterprises
Age	Mostly old	Old	Mostly young	Young	Mostly old	Old

Source: Questionnaire survey.
Note: the variables were first dichotomised (value 1 = German; high; specialised; business partnerships and capital companies; high; value 0 = Polish; low; mixed; individual enterprise; low). Subsequently, the businesses were categorised into groups (clusters) of similar characteristics by a cluster centre analysis using SPSS (Bahrenberg et al. 1992, p. 278ff).

The results show that Groups 1, 2 and 4 have the highest intensity of immaterial linkages with suppliers and buyers, and co-operate the most with other farms. The Groups 3, 5 and 6 show a strong orientation towards local material linkages and have a low intensity of local and supra-regional immaterial linkages.

   The correlation analysis shows that the large and young farms of Group 4 have the best competitive position; these farms work together with suppliers and buyers in different regions and show a combination of local cluster integration and supra-regional information flows. The importance of this strong integration in the local cluster in combination with intensive supra-regional linkages when it comes to innovations has already been identified in studies of the manufacturing and service sector (e.g. Bathelt and Glückler 2002, 213f; Fromhold-Eisebith 2000); our study on agriculture shows similar results. Probably because of their sales volume, the larger and older farms of Groups 1 and 2 also have a strong bargaining position. Their slightly weaker competitive position in comparison with the units in Group 4 might be due to the lower intensity of their supra-regional information flows, which indicates the problem of the 'lock-in' phenomenon. The weakest position is found in Groups 5 and 6, which represent the old and small farms; these units show a lack of flexibility, mobility and information, and are in danger of running out of business in the long term. Especially the poor, immobile and uniformed farms of Group 6 are highly externally dependent.

Conclusion

The study shows that rural areas appear to possess a regional agrarian cluster which covers many more activities other than agriculture, and substantially affects the regional economy. The cluster is based on material connections with the flow of goods and services; but, in addition, immaterial co-operations and information transfers are important constitutive elements of the cluster formation. And it seems that social capital – like personal contacts based on reciprocity and trust – is of great importance for the local cluster as well. But spatial proximity alone is not the only determining factor for the network formation; the level of integration is very much dependent on the behaviour of the actors and this behaviour and network integration is different depending upon the type of farm. Small units show a limited level of integration, while larger units are more intensively connected.

The spatial dimension shows that especially those farms are performing well which possess strong local links (in the sense of strong ties) in combination with supra-regional connections (in the sense of weak ties; Granovetter 1973). Ideally the best practice for farms is therefore to be embedded in local as well as in supra-regional networks. For large units with better personnel capacity, this network formation is much easier to achieve than for small units. The cluster analysis shows that not only large traditional corporations but also large younger enterprises can achieve competitiveness through immaterial linkages.

The results found in the agricultural sector are in many ways comparable to those identified for manufacturing industries. First, they show the existence of cluster-like structures in rural areas; this is important for a comprehensive spatial policy for these locations. Second, they document that, for the competitiveness of firms, combinations of different forms of linkages are important. Particularly those firms which are integrated into local and supra-regional material and immaterial linkages are successful.

References

Aydalot, P. and Keeble, D. (1986) (eds), *Millieux innovateurs en Europe*, Routledge, London.

Bahrenberg, G., Giese, E. and Nipper, J. (1992), *Statistische Methoden in der Geographie 2*, Stuttgart, Teubner.

Bathelt, H. and Glückler, J. (2002), *Wirtschaftsgeographie: Ökonomische Beziehungen in räumlicher Perspektive*, Stuttgart, Ulmer.

Camagni, R. (1991), 'Introduction: from the local "milieu" to innovation through cooperation networks', in Camagni, R. (ed.) *Innovation Networks: Spatial Perspectives*, Belhaven Press, London, pp. 1–9.

Chloupkova, J. and Svendsen G. (2003), 'Building and destroying social capital: The case of kooperative movements in Denmark and Poland', *Agriculture and Human Values*, vol. 20, pp. 241–52.

Dannenberg, P. (2006), 'Cluster-Strukturen landwirtschaftlicher Warenlketten in Ostdeutschland und Polen', Dissertation, Humboldt-Univerität zu, Berlin.

Dannenberg, P. and Kulke, E. (2005),'The importance of agrarian clusters for rural areas – results of case studies in Eastern Germany and Western Poland', *Die Erde*, vol. 136, no. 3, pp. 291–309.

Fromhold-Eisebith, M. (2000), 'Technologieregionen in Asiens Newly-Industrialized Countries', *Wirtschaftsgeographie*, Bd. 18 (Lit), Münster, Hamburg.

Gereffi, G. (1996), 'Global commodity chains: new forms of coordination and control among nations and firms in international industries', *Competition and Change*, vol. 4, pp. 427–39.

Granovetter, M. (1973), 'The strength of weak ties', *American Journal of Sociology*, vol. 78, pp. 1360–80.

Granovetter, M. (1985), 'Economic action and social structure: The problem of embeddedness', *The American Journal of Sociology*, vol. 91, no. 3, pp. 481–510.

Kulke, E. (2004), 'Wirtschaftsgeographie', Paderborn, München, Wien, Schöningh, Zürich.

Marshall, A. (1927), *Industry and Trade. A Study of Industrial Technique and Business Organization and their Influence on the Conditions of Various Classes and Nations*, Macmillan, London.

Pawlak, Z. (2004), Strukturwandel der Landwirtschaft in Westpolen in den Jahren 1991–2000 Stoll (Hrsg.), Strukturwandel in Ostdeutschland und Westpolen, *ARL Arbeitsmaterial*, vol. 311, pp. 91–102.

Pfeiffer, J. (2004), Strukturwandel der Landwirtschaft Brandenburgs', (Hrsg), Stoll. Strukturwandel in Ostdeutschland und Westpolen, *ARL Arbeitsmaterial*, vol. 311, pp. 103–15.

Porter, M.E. (1993), *Nationale Wettbewerbsvorteile*, Wien, Ueberreuther.

Storper, M. and Walker, R. (1989), *The Capitalist Imperative – Territory, Technology and Industrial Growth*, Blackwell, Oxford.

Chapter 12

Farmer Innovations in Environmental Management: New Approaches to Agricultural Sustainability?

Mairi Jay

Introduction

The aim of this chapter is to assess, critique and make a contribution towards ecological modernisation theory as applied to agriculture in a global context by examining and commenting on examples of agri-environmental stewardship in New Zealand. Ecological modernisation theory is a growing body of literature concerned with the analysis of environmentally related social change and a normative commitment to environmentally and economically sustainable change (Hertin and Berkhout 2003). While much of the early ecological modernisation literature was developed by European scholars in relation to industrial practices in Europe (Mol and Sonnenfeld 2000), more recent literature has sought to relate the theory to agriculture. This chapter will suggest that in the New Zealand case, while many producers experience pressures to improve agri-environmental performance as predicted by ecological modernisation theory, global economic processes militate against overall environmental improvements in agriculture. The undermining effects of globalisation include the fact that agri-environmental improvements by individual farmers can be offset by macro-economic drivers to intensify production while problem-specific environmental practices (such as measures to reduce pollution) do not necessarily encourage a broader stewardship ethic or motivation to reduce the array of damaging environmental effects caused by intensified agriculture.

As originally developed by German sociologist Joseph Huber (1982; 1985, as cited by Mol and Sonnenfeld 2000) ecological modernisation theory proposed that industrial practices of modernisation – such as waste reduction and energy and resource efficiencies – are an inevitable process in the development of advanced industrial societies. According to proponents, ecological modernisation does not assume that ecological sustainability and capitalist production and consumption are incompatible (Gouldson and Murphy 1996). On the contrary, a basic tenet is that, by means of environmental technologies and the transformation of modern institutions, capitalist structures can be transformed to avoid long-term environmental damage; that ecologically sustainable economic development is not only possible but necessary in order to provide for the expected increase in the human population over the next 50 years (Green et al. 2003).

With respect to agriculture, Frouws and Mol (1997) have argued that many features of the modern economy and civil society operate in ways to encourage improved environmental performance by agriculture. Pressure from consumer organisations and NGOs, the danger of losing market share to more environmentally progressive organisations, liability laws, requirements from insurance companies and cost reductions following polluter pays principles, are among many non-regulatory elements that prompt producers to improve their environmental performance. In their view, state, industry, and elements of civil society (such as consumer and environmental groups) become partners in an increasing use of voluntary agreements, environmental management systems, audit systems and eco-labelling. Evans *et al.* (2002) have suggested that the so-called 'post-productivist' characteristics observed in European agriculture, such as concern with food quality rather than quantity, the growth of sustainable farming through agri-environmental policies, and the growth of environmental regulation, are examples of ecological modernisation in the agricultural sphere. While some aspects of contemporary European social change have been prompted by the agri-environmental policies of the Common Agricultural Policy, others appear more closely related to changes in civic society. These changes, such as the demands of consumer and environmental groups, strengthen the arguments of ecological modernisation theorists that liberal capitalist societies are capable of responding to environmental concerns and moving towards environmental sustainability.

However, Porter and van der Linde (1995) suggest that market forces alone will not necessarily be sufficient to prompt firms to take the risks and costs required in innovation. They have argued that regulation is an important stimulus to environmental improvements by industry because it creates pressure for technical innovation, provides an educative role by signalling resource inefficiencies and opportunities for technological improvements, 'levels the transitional playing field' for companies that incur the costs of innovation-based practices, and creates the push to overcome compliance costs. Others, more critical of ecological modernisation theory, have argued that the theory is, 'greenwash', designed to 'defuse and deflect criticism on environmental grounds whilst corporate practices remain essentially environmentally destructive' (Gibbs 2003).

In short, ecological modernisation arguments as they apply to agriculture suggest that while the market mechanisms of capitalism may play a part in driving producers towards more sustainable forms of agriculture, they are unlikely to be sufficient on their own. In this chapter we will also argue that in an industry of multiple firms (farms), the decisions of individuals may not be enough to offset broad-scale or long term ecological effects of intensive production.

The empirical evidence for the chapter comes from a variety of sources. It includes a participant observation study of 5 dairy farm families over a 9 month period, a telephone survey of 130 dairy farmers in the Waikato region of New Zealand, extended person-to-person interviews with farmers (dairy and sheep and beef) and a thorough examination of farmer magazines and agricultural reports. Although the empirical research was primarily focused on dairy farmers, understanding of the broader agricultural context of dairying entailed wide reading and interviews with key informants.

Farmer Attitudes and Values in the Uptake of Agri-Environmental Measures

British studies have shown that the response of farmers to environmental protection schemes is varied and complex. Although productivist attitudes, values and practices predominate among farmers in most parts of Britain, a significant minority are supportive of environmental schemes, and an even larger proportion are prepared to use them as part of their farm management practices (Carr and Tait 1991; Morris and Potter 1995; Walford 2002). Farmer involvement varies by region and with the type of farms, the attitude of farmers and the age and social circumstances of farmers (for example, whether they expect a son or daughter to inherit the farm).

The importance of farmer attitudes and values for conservation, and the degree of individual variation between farmers, has been well documented (Battershill and Gilg 1997; Macdonald and Johnson 2000; Potter and Gasson 1988). It is clear that farmer attitudes and values are influenced by discourses within the wider society. A number of studies have shown changes in conservation behaviour by farmers over the 1980s and 1990s as British consumers expressed more and more anti-farming feeling in the wake of issues such as Bovine Spongiform Encephalitis (BSE or 'Mad Cow' disease), damage to the countryside from farming practices, and the cost of farm subsidies (Macdonald and Johnson 2000; Ward and Lowe 1994).

As with the British literature, New Zealand studies have found a complexity of attitudes by farmers towards environmental issues. While a majority appear to be driven firstly by concern for production and profit (Parminter and Perkins 1997; Rauniyar and Parker 1998; Underwood and Ripley 2000), environmental stewardship goals are also important (Jay 2005; 2007; Fairweather and Keating 1994; Parminter and Perkins 1997; Parminter *et al.* 1998). The studies have noted that there are farmers throughout the country who are prepared to set aside areas of native vegetation for conservation of habitat and to undertake sustainable land management measures such as planting shelter belts and erosion control (Rhodes *et al.* 2000).

Motivations for such activities include practical reasons (e.g. a desire to prevent stock losses in water courses), but also aesthetic and heritage reasons. Non-utilitarian motives may be influenced by personal factors such as age, education, stage of the family cycle, or the length of time a farm has been in the family, and expectations of succession. The views of the wider community also appear to influence a farmer's attitudes.

Economic constraints to the adoption of sustainable farming practices include low income, high debt, an ownership structure which limits the farmer's freedom to make management decisions and availability of labour. While financial constraints are often identified as a barrier to environmental management, there is evidence that many farmers expend time, effort and resources on conservation management despite financial constraints (Bradshaw *et al.* 1998). However for farmers struggling to survive, issues of long-term environmental sustainability may be pushed aside in favour of immediate financial needs (Rauniyar and Parker 1998; Rhodes *et al.* 2000; Underwood and Ripley 2000)

In terms of ecological modernisation theory, these examples from the literature suggest that wider social and economic pressures influence farmers to adopt environmental practices. The uptake of agri-environmental measures is not an automatic process, while the effects of uptake (whether long-term or short-term) depend to a significant degree on the management values and objectives of the farmer.

The Nature and Scale of New Zealand Agriculture

Pastoral agriculture is the main type of farming in New Zealand. Sheep and beef farming are the most important in terms of land area but dairying is the most important in economic terms. While dairy products contribute 19 percent of the country's exports, meat and meat products contribute 14.5 percent (Statistics New Zealand). Mainstream dairy farming involves all the characteristics described by Bowler (1992) for industrialised agriculture in western societies. It is highly specialised, highly dependent on fossil fuels and fertiliser, and highly subject to industrial modes of food production. Most farms are subordinate to food manufacturing and marketing structures and the policies of industrial agribusiness companies. Intensive practices have resulted in widespread destruction of native habitat and biodiversity (Taylor and Smith 1997); widespread soil erosion and land degradation (Taylor and Smith 1997), widespread contamination and pollution of surface and groundwater (Larned *et al.* 2005; PCE 2005; Snelder and Scarsbrook 2005) and greenhouse gas emissions[1] (PCE 2005).

A critical difference between New Zealand and European agriculture is the absence of agricultural subsidies. New Zealand farmers receive no state subsidies and are fully exposed to global market competition. The scale and intensity of pastoral farming is driven by global economic circumstances that influence the industry as a marketing and manufacturing enterprise. Farmers are subject to fluctuations in the value of the New Zealand dollar, feel vulnerable to changes in market demand and price fluctuations and regard production efficiencies as the best way to survive in the global marketplace.

The Dairy Industry Response

The New Zealand dairy industry is dominated by a single dairy company, Fonterra Co-operative Group Ltd, with some 11,000 farmer shareholders[2] (see Stringer *et al.*, Chapter 15). The milk produced by farmers is processed into either commodities such as milk powder, butter, cheese or into value-add consumer products or specialty ingredients. As a food product, marketing image and branding are important for both the commodity and value-add parts of the business even when they are sold into lower-income countries. The company strategy aims 'for global dairy industry leadership' by, among other things, remaining the 'lowest cost supplier' of commodity dairy products, the world's leading specialty milk components innovator and solutions

---

1 Domestic ruminants such as sheep and cattle are a major source of methane, a potent greenhouse gas.

2 The other two dairy companies are Tatua Co-operative Dairy Co., with approximately 140 shareholders (farm milk suppliers) and Westland Co-operative Dairy Co. with 345 shareholders.

provider, consumer nutritional milks marketer, and dairy marketer to foodservice providers (Fonterra 2006). In a recent annual report the company aims 'to maintain low-cost production structures while continuing to grow milk supply by an average three per cent year-on-year' (Fonterra 2005, 08).

A cornerstone of the industry's environmental package is the Clean Streams Accord, an agreement signed by Fonterra and representatives of regional and central government (Fonterra *et al.* 2003) (see Blackett and Le Heron, Chapter 7). The agreement is a response by the industry to domestic criticism from environmental groups and central government, and to competitive threats from overseas producers (Jay 2007). It requires Fonterra milk suppliers to prevent stock access to lakes, rivers and streams, comply with regional government regulations concerning dairy shed effluent disposal and undertake nutrient management systems. An assessment of environmental performance becomes part of the terms and conditions of supply that each farmer holds with Fonterra. Regional Councils are to monitor compliance with effluent disposal practices and Fonterra undertakes regular audits of farmers in relation to nutrient management.

Although it is early to assess the long-term effects of the programme there is evidence to indicate that a majority farmers have complied with the letter of the accord but not necessarily the spirit (MfE 2007). Figures 12.1 and 12.2 show that the message to protect streams from effluent and soil erosion has not reached some dairy farmers. Figure 12.1 shows a stream that has been fenced off as required by the Clean Streams Accord, but close against the stream fence is a heavily trampled paddock, giving little protection against soil and effluent run-off. Figure 12.2 shows a grazed hillside scarred by erosion with soil slumping into the valley bottom.

Figure 12.1   A dairy farm subject to the Clean Streams Accord

Photo: M. Jay, June 2006.

Figure 12.2   Dairy cows grazing a steep, erosion-prone hillside

Photo: M. Jay, June 2006.

The Clean Streams Accord is potentially an important way to persuade farmers to protect waterways from pollution, but raises questions of equity and enforcement. Farmers are expected to carry the cost of fencing regardless of economic or physical circumstances; farmers with multiple streams through their property will be less fortunate than those with few or no streams; farmers in physically less favoured areas will find the cost more onerous than those in more favoured regions. Equally, there is doubt that the company will be willing to pursue rigorous enforcement if the proportion of farmers who do not comply is high, or if non-compliers include farmer-shareholders who are politically influential. There is room for debate about what is a 'stream' and about whether protection of surface water (as opposed to groundwater) is sufficient to offset the effects of dairy effluent on paddocks.

Farm Environment Awards: A Farmer-based Alternative

The Farm Environment Awards began in 1993 as a joint initiative between farmers, conservationists and representatives of regional government in the Waikato region, one of the most intensively farmed areas of New Zealand. In 2004 it became a nation-wide programme directed by an independent non-profit trust (NZFEA 2007). The ten trust members are mostly farmers. The principal objective of the trust is 'the advancement, education, assistance and promotion of sustainable environmental management of land and other natural resources on farms within New Zealand' (NZFEA 2007). It aims to develop and expand its system of farm environment awards throughout the country; establish learning networks among farmers, and encourage a 'whole-of-farm' approach to environmental management.

The award process is one of the principal means of identifying and publicising environmental best practices among farmers. Some 20 to 30 farms enter the competition each year from each of 8 regions. Public open days are held on the winning farms to showcase the practices that have won the award. Assessment includes protection of natural features, management of waterways, matching land use to land types, and energy efficiency (Ballance 2006). The process is intended to highlight excellence in land management rather than excellence in business; a farmer with a difficult farm and modest profit may rate higher in the assessment process than one with an easy farm and higher level of profit.

An important consequence of the Farm Environment Award program is the social learning that occurs and the affirmation that environmental management is a significant part of being a good farmer. Judging involves teams of three or four people from different professional backgrounds, including farming, banking, accountancy, local government and conservation. They spend up to half a day or more with the farmer, listening to what the farmer has done and why, and providing advise or encouragement. Farmers learn from judges but judges also learn from

Figure 12.3   A well-formed farm track, with all-weather surface and planted up-side bank to reduce bank erosion

Photo: M. Jay, June 2006.

farmers and from each other. The learning that occurs gets passed on at subsequent judging rounds and becomes incorporated into brochures and pamphlets published by the trust. Individuals from different professional backgrounds learn to see farms holistically and are less inclined to view farm management through the lens of their own narrow expertise.

Prize money is modest and a significant motivation for many individuals is the learning they experience through the judging process and recognition of their efforts. In the words of one farmer, it is often difficult to know whether what they are doing is 'the right thing', especially in relation to environmental qualities that are not easily visible, such as stream quality or groundwater. As the farmer explained:

> It's very easy to fence a patch of bush [i.e. native forest] because you see the benefits right away. You see the regeneration come away and very quickly all the gaps fill up. So there's a feel-good factor. But if you're trying to manage groundwater, you can't actually see it; you can't know whether you're doing the right thing.

Many of the elements of sustainable stewardship are costly and only make economic sense in the long-term. For example, one farmer explained that on his rolling-to-steep hill country farm with 250kms of farm track, tracks are an important source of soil run-off into streams. As illustrated by Figure 12.3, a soundly-developed track requires proper construction of channels, culverts and an all-weather surface. The expense of sound construction makes economic sense in the long term because of the reduced maintenance required and the greater ease of stock management. Similarly, tree planting and erosion remediation are costly in the short-term but may pay-off in the long term through fewer stock losses or improved stock production.

Although information and social learning may be an important motivation for entering the competition, another is market competition. The competition is open to any type of commercial farm – livestock, arable, deer, orchards, vineyards and horticulture. Farmers who are over-represented in the awards (relative to their number) are those who produce for high-end markets, such as venison for the German market, wine for the European and North American market, quality meat cuts for Japanese and United Kingdom supermarkets, and quality fruit and vegetables (Garland, Crequer, personal communication). For these farmers, apart from the value of social learning, an important motivation is the marketing advantage that comes if their farm is noted as a winner or runner-up of the awards.

Discussion and Conclusion

Ecological modernisation theory posits that within contemporary western societies there are socio-political, economic and technological processes which lead to beneficial environmental outcomes (Buttel 2000). We have seen that in both Europe and New Zealand, there is a possible groundswell of positive environmental attitudes by farmers, although the depth and breadth of such attitudes is variable, and does not necessarily flow through to environmental practice. While some farmers are prepared to make significant changes to improve environmental outcomes, others express positive attitudes but are driven by stronger financial priorities. How do the New Zealand agri-environmental initiatives just described weigh up as moves towards environmental sustainability and to what extent are they inherent to the socio-political, economic and technological processes suggested by the theory?

At this time, neither program has achieved large-scale, country-wide shifts in sustainable farming practice but both appear to have contributed to a raised climate of awareness among farmers and willingness among some to adopt environmental management systems. The Clean Stream Accord has drawn the attention of a majority of dairy farmers to the serious environmental consequences of surface water pollution. Before the Accord few Fonterra farmers fenced off their streams from stock access or practised nutrient budgeting. Regional government regulations were routinely ignored. Since the Accord, farmers who persistently ignore the terms of the agreement may fail to have their milk accepted by the company.

Similarly, the Farm Environment Award has fostered a whole-farm environmental ethic which encourages farmers to note the environmental impacts of all aspects of their land management. The ideal of the good farmer becomes less one who produces the most or makes the biggest profit, and more the one who manages the land most skilfully from a long-term perspective. Equally significant, the Farm Environment Awards have reinforced notions of quality rather than quantity as a market strategy. In this regard, the programme is similar to a growing number of food quality assurance and farm environmental schemes (Wharfe and Manhire 2005). Examples include meat companies that supply UK supermarkets, and environmental management programmes by the kiwifruit, pipfruit, viticulture and deer industries. In the face of competition from other southern hemisphere countries such as Australia, South Africa and Chile, these producers regard quality as an important competitive advantage.

Both programmes support the view that capitalism – in the form of market pressures – has the capacity to encourage environmental responsibility. New Zealand's dependence on exports of quality agricultural products to the wealthier markets of North America, Japan and Europe has been a significant motivator for farmers to undertake agri-environmental improvements.

However, both programmes have weaknesses from the perspective of ecological sustainability. Clean Streams encourages broad uptake by a majority of dairy farmers, but is narrow in scope, while the Farm Environment Award encourages a broad environmental awareness, but is limited in its spread. The Clean Streams Accord does little to change farmer attitudes towards environmental sustainability; it is simply regarded by many farmers as another management hurdle along with milk hygiene, animal welfare and occupational health and safety. Nor does it generate social learning about broader sustainability issues. In contrast, the Farm Environment Award encourages social learning by a wide cross-section of the farming community, but involves only a very small percentage of farmers (a few hundred among thousands). Whereas the Farm Environment procedures involve a large volunteer commitment in time and energy from the farming community, the Clean Streams Accord puts the onus on farmers to comply with a defined and limited task that is easy to monitor and cheap to administer relative to the number of farmers involved (all 11,000 of Fonterra's milk suppliers).

Perhaps most importantly, neither programme is able to reduce the effects of agricultural intensification. As indicated by Table 12.1, the intensity of dairying, as measured by the production of milksolids (fat and protein) per hectare, has increased 10 percent in the 5 years from the 2000/01 milking season to 2005/06 season. Most of the effluent from cows' lands on the paddock rather than the dairy shed (MfE 1999) and so the increase in number of cows per hectare (stocking rate) and milk solids per hectare represents an increase in effluent on paddocks. While surface water pollution is an important issue, equally important in the long term is groundwater pollution. Effluent from cows in the fields percolates into the groundwater and over time flows into lakes and streams.

Table 12.1   Change in dairy intensity, 2000/01–2005/06

Indicators of dairying intensity	2000/01	2003/04	2005/06	2000/01–05/06
				% increase
Total number of dairy cows	3,485,883	3,851,302	3,832,145	10
Average kg of milksolids per cow	310	315	325	5
Average cows per hectare	2.66	2.75	2.77	4
Average kg of milksolids per hectare	825	889	907	10

Source: LIC 2006.

As well as water pollution, dairying can cause soil erosion (Figure 12.2), soil compaction, destruction of native habitat, loss of biological diversity, and greenhouse gas emissions[3] (methane from cows).

The intensification of sheep and beef has been less marked and less rapid, but between 1991 and 2002 fertiliser use by sheep and beef farms increased between 24 percent and 28 percent (PCE 2004). At least in relation to dairying, economic forces of higher costs relative to returns mean that farmers are running to stand still. Production increases, for most, are necessary simply to offset rising costs (PCE 2004).

In conclusion, the practices of ecological modernisation, such as industry-based food and environmental quality standards, have become increasingly common within global food networks. An assessment of ecological modernisation theory in the light of the New Zealand experience suggests that there are indeed socio-political processes within a modern capitalist society that serve to generate pressures for improved agri-environmental performance. But these processes are not sufficient to off-set a global economic pressure on Fonterra to increase and intensify production and to define environmental improvement in the narrowest terms. Nor are they sufficient to create a holistic stewardship ethic that is widespread among farmers.

Unless the position of farmers relative to retailers and consumers can be rebalanced, and unless the growth machine can be redirected, we are unlikely to witness an overall improvement in the state of New Zealand's environment. So long

---

3 Methane from cows is a significant greenhouse gas.

as Fonterra aims to 'grow milk supply by an average three per cent year-on-year' (Fonterra 2005, 08), environmental degradation is virtually inevitable.

Acknowledgements

Mairi Jay would like to acknowledge Bill Garland, farmer, for his generous and honest information, Gordon Stephenson, retired dairy farmer, for his steadfast commitment to sustainable farming, and Philippa Crequer, Executive Officer, Waikato Farm Environment Trust.

References

Ballance (2006), Balance Farm Environment Awards, <http://www.ballance.co.nz/fea.html> (accessed June 2006).

Battershill, M.R.J. and Gilg, A.W. (1997), 'Socio-economic constraints and environmentally friendly farming in the Southwest of England', *Journal of Rural Studies*, vol. 13, no. 2, pp. 213–28.

Bowler, I.R. (1992), 'The industrialisation of agriculture', in Bowler, I.R. (ed.), *The Geography of Agriculture in Developed Market Economies*, Longman Scientific and Technical, Harlow, Essex, pp. 7–31.

Bradshaw, B., Cocklin, C., and Smit, B. (1998), 'Subsidy removal and farm-level stewardship in Northland', *New Zealand Geographer*, vol. 54, no. 2, pp. 12–20.

Buttel, F.H. (2000), 'Ecological modernization as social theory', *Geoforum*, vol. 31, pp. 57–65.

Carr, S. and Tait, J. (1991), 'Differences in the attitudes of farmers and conservationists and their implications', *Journal of Environmental Management*, vol. 32, pp. 281–94.

Evans, N., Morris, C and Winter, M. (2002), 'Conceptualizing agriculture: a critique of post-productivism as the new orthodoxy', *Progress in Human Geography*, vol. 26, no. 3, pp. 313–32.

Fairweather, J. and Keating, N.C. (1994), 'Goals and management styles of New Zealand farmers', *Agricultural Systems*, vol. 44, no. 2, pp. 181–200.

Fonterra (2003), About Fonterra. Internet homepage, Fonterra, Auckland, New Zealand. <http://www.fonterra.com/default.jsp> (accessed 18 August 2003).

Fonterra (2005), *To Lead in Dairy*, Annual Report 04/05, Fonterra Corporate Centre, Auckland, New Zealand.

Fonterra (2006), Home page, About Fonterra, Strategy, <http://www.fonterra.com/content/aboutfonterra/strategy/default.jsp>.

Fonterra, Ministry for the Environment, Ministry of Agriculture and Forestry, Local Government New Zealand (2003), Dairying and Clean Streams Accord, Between Fonterra Co-operative Group, Regional Councils, Ministry for the Environment and Ministry of Agriculture and Forestry, Ministry for the Environment, Wellington.

Frouws, J. and Mol, A.P.J. (1997), 'Ecological modernisation theory and agricultural reform', in De Haan, H. and N. Long (eds), *Images and Realities of Rural Life, Wageningen Perspectives on Rural Transformations*, Van Gorcum, The Netherlands, pp. 269–86.

Gibbs, D. (2003), 'Reconciling economic development and the environment', *Local Environment*, vol. 8, no. 1, pp. 3–8.

Green, K., Harvey, M. and McMeekin, A. (2003), 'Transformations in food consumption and production systems', *Journal of Environmental Policy and Planning*, vol. 5, no. 2, pp. 145–63.

Gouldson, A. and Murphy, J. (1996), 'Ecological modernization and the European Union, *Geoforum*, vol. 27, no. 1, pp. 11–21.

Hertin, J. and Berkhout, F. (2003), 'Analysing institutional strategies for environmental policy integration: the case of EU enterprise policy', *Journal of Environmental Policy and Planning*, vol. 5, no. 1, pp. 39–56.

Huber, J. (1982), *Die verlorene Unschuld der Ökologie. Neue Technologien und Superindustriellen Entwicklung*, Fisher Verlag, Frankfurt am Main.

Huber, J. (1985), *Die Regenbogengesellschaft, Ökologie und Sozialpolitik*, Fisher Verlag, Frankfurt am Main.

Jay, M. (2005), 'Remnants of the Waikato: native forest survival in a production landscape', *New Zealand Geographer*, vol. 61, no. 1, pp. 14–28.

Jay, M. (2007), 'The political economy of a productivist agriculture: New Zealand dairy discources', *Food Policy*, vol. 32, no. 2, pp. 266–79.

Larned, S., Scarsbrook, M., Snelder, T. and Norton, N. (2005), *Nationwide and Regional State and Trends in River Water Quality 1996–2002*, Prepared for the Ministry for the Environment by the National Institute of Water and Atmospheric Research Ltd, Ministry for the Environment, Wellington, New Zealand.

Livestock Improvement Corporation (2006), *Dairy Statistics 2005–2006*, Livestock Improvement Corporation, Hamilton, New Zealand.

Macdonald, D.W. and Johnson, P.J. (2000), 'Farmers and the custody of the countryside: trends in loss and conservation of non-productive habitats 1981– 1998', *Biological Conservation*, vol. 94, pp. 221–34.

MfE (Ministry for the Environment) (1999), Resource Management Act Practice and Performance: Are Desired Environmental Outcomes Being Achieved?, Ministry for the Environment, Wellington, New Zealand.

MfE (2007), *The Dairying and Clean Streams Accord: Snapshot of Progress – 2005/2006*, Ministry for the Environment, Wellington, New Zealand, <http:// www.mfe.govt.nz/publications/land/dairying-clean-streams-accord-snapshot-mar07/dairy-clean-streams-accord-snapshot-mar07.html>.

Mol, P.J.A. and Sonnenfeld, D.A. (2000), 'Introduction' in Mol, A.P.J. and Sonnenfeld, D.A. (eds), *Ecological Modernisation Around the World, Perspectives and Critical Debates*, Frank Cass, London, pp. 3–16.

Morris, C. and Potter, C. (1995), 'Recruiting the new conservationists', *Journal of Rural Studies*, vol. 11, pp. 51–63.

Murphy, J. (2000), Ecological modernisation, Editorial, *Geoforum*, vol. 31, pp. 1–8.

NZFEA (2007), New Zealand Farm Environment Award Trust, <http://www. nzfeatrust.org.nz/content/29/default.aspx>.

PCE (Parliamentary Commissioner for the Environment) (2004), *Growing for Good, Intensive Farming, Sustainability and New Zealand's Environment*, Parliamentary Commissioner for the Environment, Wellington, New Zealand.

Parminter, T. and Perkins, A.M.L. (1997), 'Applying an understanding of farmers' values and goals to their farming styles', *New Zealand Grassland Association*, vol. 59, pp. 107–11.

Parminter, T.G., Tarbotton, I.S. and Kokich, C. (1998), 'A study of farmer attitudes towards riparian management practices', *New Zealand Grassland Association*, vol. 60, pp. 255–8.

Porter, M.E. and van der Linde, C. (1995), 'Toward a new conception of the environment-competititveness relationship', *Journal of Economic Perspectives*, vol. 9, no. 4, pp. 97–118.

Potter, C. and Gasson, R. (1988), 'Farmer participation in voluntary land diversion schemes', *Journal of Rural Studies*, vol. 6, pp. 1–7.

Rauniyar, G.P. and Parker, W.J. (1998), Constraints to Farm Level Adoption of New Sustainable Technologies and Management Practices in New Zealand Pastoral Agriculture, Department of Agribusiness and Resource Management, Massey University, MAF Policy Technical Paper 98/3, Ministry of Agriculture, Wellington, New Zealand.

Rhodes, T., Willis, B. and Smith, W. (2000), Impediments to Optimising the Economic and Environmental Performance of Agriculture, Vol. 1: A Study of Issues Affecting North Island Hill Country Farmers, Technical Paper 2000/17, MAF Policy, Ministry of Agriculture and Forestry, Wellington, New Zealand.

Snelder, T., and Scarsbrook, M. (2005), Spatial Patterns in State and Trends of Environmental Quality in New Zealand Rivers: An Analysis for State of Environment Reporting, prepared for the Ministry for the Environment by the National Institute of Water and Atmospheric Research Ltd, Ministry for the Environment, Wellington, New Zealand.

Statistics New Zealand (2004), *New Zealand Official Yearbook*, Government Printer, Wellington, New Zealand.

Taylor, R. and Smith, I. (1997), *State of New Zealand's Environment 1997*, Ministry for the Environment, Wellington, New Zealand.

Underwood, R. and Ripley, J. (2000), *Impediments to Optimising the Economic and Environmental Performance of Agriculture*, vol 2: *Review of Literature*. MAF Policy, Ministry of Agriculture and Forestry, Wellington, New Zealand.

Walford, N. (2002), 'Agricultural adjustment: adoption of and adaptation to policy reform measures by large-scale commercial farmers', *Land Use Policy*, vol. 19, pp. 243–57.

Ward, N. and Lowe, P. (1994), 'Shifting values in agriculture: the farm family and pollution regulation', *Journal of Rural Studies*, vol. 10, no. 2, pp. 173–84.

Wharfe, L. and Manhire, J. (2005), The SAMsn Initiative: Advancing Sustainable Management Systems in Agriculture and Horticulture, An Analysis of International and New Zealand Programmes and their Contribution to Sustainability, The AgriBusiness Group, Consultants, for the Sustainable Management Systems Network (SAMsn), Wellington, New Zealand.

Personal Communication

William (Bill) G. Garland, sheep and beef farmer, ex-judge for the Farm Environment
  Awards.
Gordon Stephenson, dairy farmer and ex-Chairman of Farm Environment Awards.
Philippa Crequer, Secretary, Farm Environment Award for Waikato region.

Chapter 13

The Region as Organisation: Differentiation and Collectivity in Bordeaux, Napa, and Chianti Classico

Jerry Patchell

Introduction

To compete with the increasing integration of the supply chain by an oligopoly dyad of multinational wine/beverage corporations and supermarkets, winegrowers have developed similar systems of organisation. A growing literature supports this perspective. I add to it by comparing practices in three leading wine regions and by incorporating the need for differentiation within regions, both among different territories and individual winegrowers. I focus on estate winegrowers because they compel differentiation and therefore shape territorial and regional structures.

Why Bordeaux, Napa and Chianti Classico? Their reputation precedes them, but these regions were built with great effort. Bordeaux is generator or propagator of many innovations taken for granted in the wine industry: the chateau or estate, appellations, *terroir*, estate bottling, differentiated futures markets. It's also undergoing a transformation from bulk to estate wine sales. Napa spearheads the diffusion of chateau and *terroir* to the New World, innovating to do so. Chianti Classico also represents the diffusion of estates, but with a twist. To establish identity within the internationalisation of wine styles it is re-establishing territorial unity and control. The three regions demonstrate what type of markets, institutions and economies of scale can be constructed to build regional organisation and how it is shaped by the tensions between collectivity and differentiation.

Oligopolisation – Versus Regional Organisation

Oligopolisation of the wine industry has two elements joined by the power of distribution and marketing. First, multinational producer-distributor firms are increasing scale and scope through acquisitions to exploit the globe's vineyards and wine markets (Coelho and Rastoin 2004). The multinational acquisition and marketing strategies of European, North American and Australian firms belie any simple New/Old world dichotomisation. Second, the retailing sector, particularly supermarkets, is oligopolistic in all developed country markets (Wrigley 2001) and increasingly dominates wine sales. Supermarket influence is strongest in Northern

European countries (e.g. UK 50 percent, Netherlands 62 percent) that have little domestic production and are important markets. Oligopoly is also important in North American food services because of chain restaurants. The interplay within the oligopoly dyad is animated by a struggle over margins and industry analyses describe this process as incipient.

Distribution power reduces suppliers, demands more in terms of services and range of products or removes them with own brands. Suppliers must invest in innovation, distribution, and marketing of a range of brands spanning price points from basic to super premium. The scale advantages brought to bear by the giant retailers and producer/distributors include: leverage over fragmented suppliers (grape growers and estates); the ability to scale production up and down and force costs of fluctuations of suppliers or distributors; integrated and authoritarian quality control; R&D that provides advantages for internal production and leverage over suppliers; mass advertising and umbrella branding; leverage/control over shelf space; own distribution channels/outlets; trademark protection and legal/regulatory depth; political clout.

To some the number of independent producers and inherent potential for heterogeneity in the global wine industry provide comfort that a win-win situation between regions and brands can prevail. Others claim backward integration is necessary, particularly in the old world, or that the rising numbers of independent winegrowers are a fringe of niche specialists dependent on local or ultra premium markets existing in the shadow of large generalists (Swaminathan 2001). Indeed the volumes produced by the relative small numbers of specialists in the U.S. and Australia (despite dramatic recent growth) suggest the limited role. Multinational producer/distributors, however, aren't simply interested in brands, but gain their best margins mimicking small producer strategies in their ultra-premium and icon ranges (Shelton 2001; Swaminathan 2001). They capitalise for their brands by developing estates in leading regions or buying leading estates, capturing associations that often have taken centuries to build.

Wine regions predate the oligopolies by centuries, but it is only within the last half century that they have developed as platforms for horizontal integration among independent producers. Bordeaux, Napa, and Chianti Classico have long established regional reputations and their acreage and volumes have increased, but more importantly differentiation by grape quality, *terroir* and, most autonomously, estate has multiplied. Counter-intuitively the value added of independent winegrowers increased. From the 1960s estate bottling and self-branding internally diffused to become the dominant form of production today. Concomitant with internal diversification has been the elaboration of various institutions for collective action and integration. These institutions and the markets, and economies of scale and scope offered to winegrowers are illustrated in Table 13.1. They are generated by the mutual need for collective and individual reputations, but bridging these needs generates internal tensions.

Table 13.1   Functions and regional institutions of Bordeaux Napa and Chianti Classico

Functions	Institutions			
	Bordeaux	**Napa**	**Chianti Classico**	**External**
Markets	La Place (Union des Grand Crus; Cercle de Rive Droite; Syndicats) Vinexpo (Chamber of Commerce)	Premier NV; NV Wine Auction (NVV)	Yearly Tastings (Consorzio)	State/ National/ EU/ International trade boards; regulations
Production/ Price Control	INAO/ Syndicats; CIVB		Consorzio, Sienna, Firenze, Tuscany	State/ National; Winegrowers; Chambers
R&D, Training	Syndicats, CIVB, Universities, Chamber of Agriculture, Others	NVV, Universities, County	Consorzio, Universities	State/ National; Winegrowers
Quality Control	Syndicats/ INAO, CIVB		Consorzio, Chambers	State/National; Winegrowers
Promotion	CIVB, Syndicats, Academie du Vin, Vinous Brotherhoods, Tourism offices, Vigneron Independents	NVV, AVA Associations	Consorzio, MTVT, Città del Vino, Sienna, Firenze, Tuscany	State/National/ International trade boards; marketing offices; wine associations
Fraud Control	Syndicats/INAO	NVV, NVGA, County, State	Consorzio	State/National/ EU/ International
Distribution Politics	CIVB, Winegrower & Négociant Syndicats, Coops (FCVA)	Free the Grapes, NVV, Family Winegrowers, Wine Institute, NVGA	Consorzio, Coops	National/States; Farmer-merchant unions/ Federations
Regulatory Politics/ Information	CIVB	NVV, NVGGA, Wine Institute	Consorzio	National/States; Farmer-merchant unions/ Federations
Social-Environmental responsibility	CIVB, Region of Acquitaine, MSA	NVV, NVGGA, County, NGOs	Consorzio, Tuscany, Siena, Firenze	National/ State; NGOs

Note: AVA (American Viticultural Area); Ch. Of C. (Chamber of Commerce); CIVB (Conseil Interprofessionnel du Vin de Bordeaux); FCVA (Fédération des Caves Coopératives Vinicoles d'Aquitaine); FGVB (Federation de Grand Vins du Bordeaux); INAO (Institut National des Appellations d'Origine Conrolrôlée; MSA (Mutualité Sociale Agricole); MTVT (Movimento Del Turismo Del Vino Toscana); NVV (Napa Valley Vintners); NVGGA (Napa Valley Grapegrowers Association).

Regional reputation draws winegrowers into collective action because it reduces information costs for consumers (Arfini 2000; Giraud-Héraud *et al.* 1998) in the same way brands identify product quality in distanced markets. Indeed the integration of the brand and place of origin concepts is increasingly accepted (Orth 2005) and the regional reputation's value to consumers well established (Schamel and Anderson 2003). Appellation of origin or geographical indicators can also provide intellectual property rights at national and international levels, giving winegrowers a legally defensible monopoly power (Moran 1993). These regional rights, however, are more ambiguous and receive less respect than corporate trademarks (Escudero 2001).

Reputation must be maintained over the long term to be effective (Tirole 1996), and indeed a benefit of the region is provision of a venue for long-term reputations and therefore the extraction of higher premiums than short-term performance (Landon and Smith 1997). European protected designation of origin (PDO) and protected geographical indication (PGI) literature provides examples of how a region can be managed by a collective: quality control; control of volumes; contractual governance in the supply chain; promotion; and R&D and training (Arfini 2000; Barjolle and Sylvander 2000; Giraud-Héraud *et al.* 1998). Examples of regional coordination for R&D, mutual assistance, quality control, etc. have also been noted in Australia, Chile, and Canada. The broader territorialisation processes are also emphasised (Lewis *et al.* 2002). Gade describes Cassis as 'overwhelmingly the collective will of the *syndicat*' (2004, 863), and the commingling of winegrowers with municipal government and municipal services. Focused on tourism, wine's fostering of multifunctional agriculture evokes territorial collective action (Belletti *et al.* 2002).

Regional reputation therefore provides incentives for collectivity and generates supporting organisations, however, differentiation produces significant tensions. For many winegrowers the value added by regional reputation is not enough. They seek to produce the market power that will acquire the value normally captured farther down the value chain. Estate bottling represents the most effective means devised to lessen the price-taker role of the agriculturalist. Whether they choose this strategy for love or money (Scott and Podolny 2002), the choice creates a powerful dynamic within any collectivity. Differentiation needs to be muted, however, in order to maintain the information cost reductions of regional reputations required by consumers (Giraud-Héraud *et al.* 1998). Estate winegrowers overcome this contradiction by diversifying their product within a coherent regional style. Thus originating in France, and increasingly elsewhere, regional identity (*typicité*) is paired with the differentiation of the unique human and physical inputs of an estate known as *terroir*.

The tensions produced by regional organisation begin with the limitations differentiation puts on vertical integration. Wine regions exhibit variations of seller-buyer relationships between grapegrower/winegrowers, cooperatives, merchant-blenders, and merchants (of estate-bottled wine). They can be competitors or collaborators, or both, but there is always potential for opportunism. This strategising can lead to under-investments in viticulture, viniculture, and marketing, especially when the conditions of agriculture are considered. Giraud-Héraud *et al.* (2000) suggest amelioration by providing market and transaction information for both sides and developing stabilising contractual arrangements administered by the *interprofession* (associations of buyers and sellers). These mechanisms mediate grape and bulk wine sales, but differentiation by *terroir* and estate complicates the relationship. When winegrowers take control over differentiation merchants have neither capacity nor incentive to represent dozens of small chateaux, rather they are likely to free ride on the reputation of chateau and appellation.

If regional reputation is important, then it makes sense to expand generic advertising, including making it compulsory for producers in an industry or region. It is, however, difficult to connect general benefit with individual returns, particularly when producers are trying to establish their own brands. Crespi and Marette (2002) found that generic advertising can hurt differentiation and Marette (2003) claims that while AOCs can benefit from combined efforts in advertising, this potential may be undermined by regional differences. While small-scale producers may feel compelled to join such generic schemes despite their ambiguous returns, the promotional scale of MNCs allows them to opt out, while retaining benefits of reputation.

The tension between generic and individual brands relates to a weakness of regional reputation, even when protected by appellation. Corporate trademarks are owned privately, are transferable, and can be authoritatively protected and supported. On the other hand, the appellation is a non-transferable public good, shared by producers in a region and requiring national regulations to enable producers to put even limited controls on rights to the appellation. Because of this weakness, and the literature extolling brands, some winegrowers may put their faith in a brand that, in addition to selling, can be sold. The inability to transfer the regional trademark is partially compensated for by the transferability of land.

A further ramification of the lack of property rights is less incentive to protect the trademark. However, building the capacity to protect reputation is essential to maintaining an organisation (Tirole 1996). Regional organisation, therefore, often requires quality control: the setting of minimum standards, monitoring and enforcement (Winfree and McCluskey 2005). In a region of independent principles this entails self-governance to create standards and regulations and self-monitoring to prevent freeriders (Ostrom 1990).

The most salient point of comparison between regions and corporate organisation (other than the producers' independence) is that where corporations may make corporate social responsibility and location volitional, to winegrowers concerned about regional reputation and tourism, contributing to the environmental and social fabric is a necessity (Belletti *et al.* 2002; Gade 2004; Roudié 1991). Winegrowers have to be concerned about negative and positive externalities.

Balancing Collectivity and Differentiation

The tensions described in the previous section are related to the institutions of the three regions on Table 13.2. In the following, I briefly comment on how the tensions are dealt with in each region, and highlight one example. However, as in any complex and semi-integrated structure there is interdependence between the functions of the various institutions. Almost all the tensions shown in Table 13.2 are evident in each region, but the means of resolving them differs in kind or degree.

Table 13.2   Tension between collectivity and differentiation

	Bordeaux	Napa	Chianti Classico
Collectivity: Regional-territorial	QC (upstream/ downstream) Production control Promotion R&D Politics Territorial environment	Promotion R&D Politics Social responsibility Territorial environment	QC (upstream) Production control Promotion R&D Politics Territorial environment
Tensions	⟵	Sell/buy relations Promotion returns Cohesion/recognition Quality disparities Externality sharing	⟶
Differentiation	Appellations Classifications En primeur Coop chateau labels Cybersales Internal/external groups	Appellations Vineyard designates, Proprietary labels Custom crush Napa premier/auction Cybersales Internal/external	Riserva Village groups Cybersales Internal/external

Sell-Buy Relations

Mediation of sell-buy relations evolved to deal with commodity transactions of grape or bulk wine producers and wineries or merchant blenders. Bordeaux's CIVB is the only institution developed specifically for this relation. With legal mandate to extract fees and to collect information, at one time it managed prices and volumes, but now focuses on promoting demand and eliminating price information asymmetries. Similarly the NVGGA helped commodity producers by pushing the State to extract and publish prices and it educated vintners on grape grower costs. Chianti Classico, set a floor price for grapes in some years, but otherwise doesn't manage transactions.

A diversity of contracts specifying duration and grape qualities now enable producers to differentiate and negotiate with buyers on that basis. Napa grapegrowers introduced bottle pricing to get a fairer share of retail prices and a few constructed custom crush facilities for buyers to make wine. Cooperatives in Chianti Classico and Bordeaux reveal the continuing importance of capturing value through collective integration. The pervasive tool to capture value, however, is the winegrowing estate. It enables escape from commodity prices because bottles need not be sold on short-term markets like grapes or bulk wine. The challenge is to market this diversity of differentiation. Napa has organised (Free the Grapes) to open up direct sales in state-by-state wholesale oligopolies, while the Consorzio members collaborate to bring tourists to the region and its B & Bs.

Bordeaux has to provide marketing opportunities for 5000 chateaux. To that end the CIVB's fostered fall wine fairs in France's supermarket oligopsony and along with the Chamber of Commerce developed Vinexpo (largest trade fair). Perhaps the opportunities and challenges of estate-based differentiation are best demonstrated by partial conversion of 400 regional merchants to a global sales force for a myriad of chateaux. *La place de Bordeaux* originates in the centuries old practice of merchants negotiating for and buying wine in the spring (*en primeur*) and then transporting it to their cellars for aging and blending. When chateau bottling diffused, chateaux began to offer tastings to merchants, importers, and wine critics in the spring thereby enabling them to set prices, sell wine before bottling, and improve cash flow tremendously. No longer hostage to merchants, chateaux can reserve or withdraw allotments year to year to encourage loyalty, stimulate sales efforts, and stagger sales of allotments to capture upward speculation in price. Importantly, chateaux use Bordeaux merchants to maintain the regionally based distribution system. *La place* exemplifies how futures markets, rather than fixing commodity prices, can be used for differentiated agricultural goods.

The evolution of *la place* was fostered by the Union des Grand Crus, an association of about 120 top chateaux organising tastings, accommodation and entertainment for journalists, critics, and merchants. The Cercle de Rive Droit (top right bank chateaux) and some *Syndicats* also operate *en primeur* events drawing 5000 people in 2006. The shift in power to chateaux brings claims of gouging, disloyalty, and inability to serve customers properly from merchants, but the chateaux need a diversity of channels for global distribution of small volumes. Only 200 chateaux use the *en primeur* system with leverage, but the system serves less-known estates because merchants offer a complete range of chateaux and because chateaux marketing efforts and territorial reputations provide leverage with merchants. The NVV has attempted to establish a futures market through its charity auction and Premiere Napa Valley® events.

Promotion Returns

Tension between collective and individual promotion aggravated the split between the technical Consorzio Chianti Classico and the marketing Consorzio del Marchio Storico. The largest producers, Antinori and Frescobaldi, pulled out of both, saving the fees for their own promotion (Belfrage 2001). Other famous small to medium sized estates dropped out of the marketing consorzio because they were unhappy with collective promotion or with its style. One the other hand tourism territorialisation by the consorzio and other organisations is very important to the majority of small and medium estates.

The CIVB spends much of its budget on advertising, and doubled its fees to counter New World brands. Yet despite recognition that 'Bordeaux' opens doors, chateaux complain about generic advertising that can't be tied to individual results, is dated and that reinforces an elitist image. The more famous territories/appellations in Bordeaux even resist putting Grand Vin du Bordeaux on the bottle. The CIVB has addressed these issues by allocating more advertising to the appellation families and emphasising the *petits chateaux*.

The NVV does little overt advertising, but facilitates and leverages media interest in the valley and runs events where wineries can draw attention to themselves. On the other hand, in Napa there is more tension between the appellation and the brand than in Europe. In the 1960s prestigious estates were diluted after corporations bought them to use their brand names for volume production. Today's large beverage companies (some with 800 ha) are posing other challenges. After taking over Mondavi, Constellation immediately moved to increase production the same premium brands with more volume. Another tactic is to place wines from other regions within product ranges named after a Napa leader (e.g. Beringer's). Medium-sized Napa origin producers also follow this strategy, developing a range of wines outside the valley, and selling them to unwitting tourists or those unwilling to pay Napa prices (e.g. Niebaum-Coppolla). Producers will likely try to limit the adverse impacts of this designed ambiguity because the premium wines offer higher margins and because they provide a halo affect for their range. Yet, the temptation to amplify a brand is ever-present, because brands are so powerful. Dan Duckhorn, who built one of California's more successful medium sized operations from Napa cachet claims brand equity is far more important than appellation. On the other hand, Augustin Huneeus, who built the global reputation of Concha y Toro argues that brands will never divert fine wine consumers interested in place and discovery. A moot point for the 60 percent of Napa winemakers making less than 10,000 cases – they need the appellation.

Coherence and Recognition

Grape varieties, chateaux, the shape of the bottle, the landscape and climate, Bordeaux tries to project a coherent image, but it has also evolved mechanisms to allow for territories, groups and individuals to differentiate themselves. Perhaps, the outstanding differentiation mechanisms are the classification systems set up by the winegrowers to allow a superior few to garner greater profits. Recognition through these many vehicles, however, is a double-edged sword. Consumer confusion with 57 appellations testifies to the price paid in coherence for differentiation.

Napa's differentiation challenge in is not so extreme as Bordeaux, but with wineries increasing from a few dozen in the 1970s to over 300 today, differentiation pressure increases. The sub-AVAs offer substantial differentiation because they are small, only a few have more than a dozen or two wineries. Although supporting this differentiation, the NVV also obtained a law of 'conjunction' giving Napa pre-eminence over other AVAs on labels. Increasing dominance of cabernet sauvignon in the main valley and pinot noir and chardonnay in cooler Carneros also provide appellation coherence. And to defend their exclusive rights to this signal the NVV and the NVGGA have obtained *sui generis* laws affording the protection of the AOC system.

While Napa and Bordeaux accommodate differentiation, Chianti Classico fights for coherence on two interrelated fronts. The territory has been fighting for its identity and autonomy since the 1930s when the first Italian appellation system designated 'chianti' as a wine style that could be used outside the territory and required that governance over the style be shared by different territories. After negotiations with many governments and repeated revisions of the DOC(G) laws, Chianti Classico was granted the right to implement unique regulations in 1996. Then in 2005 all consorzii were given the power to impose regulations on appellation users. The transformations in government stance were in-part compelled by producers' rebellion against the DOC(G) regulations. In the 1980–90s 'super-Tuscan' producers eschewed the uncompetitive chianti formula, and sold their best wines as IGT or table wine.

The super-Tuscans helped win the war for autonomy, but inflicted collateral damage on the reputation of the appellation, especially by removing some locomotives. The *consorzio* has worked to recapture these locomotives and rebuild coherence, and is assisted by a few incentives. Winegrowers produce predominately sangiovese and chianti classico and therefore have an interest in the reputation of the appellation. They believe linking sangiovese, territory, and chianti classico distinguishes them in an international market dominated by Bordeaux and Burgundy varieties. Furthermore, sangiovese, chianti and variety are important to them personally. Nor is coherence overly threatened by differentiation. The territory has not produced a classification or sub-appellation because winegrowers don't think they apply to the landscape and because they fear divisiveness. Associative differentiation does occur, through villages and activities like wine tastings, and one small group now calls itself super-chianti. For the most part, however, differentiation is done on an individual production and marketing basis.

Quality Disparities

Napa is not unaware of the need to overcome quality disparities as there have been problems with underperformers. However, the AVA system doesn't allow for quality control (QC) and Napa can rely on competition and its attractiveness to capital to elevate standards. Bordeaux and Chianti Classico, however, needed to convert peasant-based agriculture into territories built on estate winery reputations. Sharecropping dominated Chianti until the 1960s, leaving a legacy of uncompetitive vine varieties and planting configurations. Overcoming this legacy generated much divisiveness between and within the two consorzii, particularly as many winegrowers couldn't or wouldn't keep up with constantly upgraded regulations. The amalgamated consorzio now has legal authority to force improvements. Bordeaux's QC challenge is much greater because of its size and the limitations of its common QC system.

In 1960 Bordeaux was fragmented among 47,000 winegrowers, many who wanted to 'make the vines piss'. Only 10,000 remain today, but Bordeaux must maintain standards for fine wine in a region that can produce 7 million hl and is composed of territories with disparate quality aspirations. Half that volume results from expansion in the last thirty years: 6000 ha of new land, conversion from non-AOC white, and yield increases (Roudié 1988). Much expansion incorporated vines, spacing, facilities, and attitudes differing from the estate wineries. Expansion allowed many quality estates to come to the fore, but also created opportunities for free riding and necessity to protect reputations.

The territorial *syndicats* are the foundation of Bordeaux's multi-level QC system. They determine, monitor, and enforce regulations within an AOC QC framework that requires basic procedures to be followed. The same procedures accused of allowing collusion in testing and discouraging innovation by enforcing standardisation (typicité) (César 2002). A certification process that takes a representative blend of all vats as a sample for testing, and allows for later blending of vats into superior and inferior wines, all with AOC labels, is the unsound foundation of the system. The system does, however, allow for stricter controls and differentiation – dependent on the volition and consensus making capacities of territorial winegrowers to compel differentiation. Saint Emilion, for example, is a compact territory of 800 estates that built its reputation by creating classifications and appellations that foster competition and innovation. The 6,500 members of the syndicat of Bordeaux/Bordeaux Supérieur can reap economies of scale for training, R&D, and testing, but varying *terroir*, financial and technical capacities, and different standards between estates and bulk wine producers challenge cohesive upgrading.

At the regional level, the CIVB samples bottles in distribution for testing and grading. The results are given to merchants and winegrower *syndicats* to clarify accountability. The INAO, worried about the reputation of the AOC system suggested creating an AOC d'Excellence for another *terroir* based distinction. Bordeaux rejected this idea because of its complicated appellation and classification systems and for imposing on *syndicat* autonomy. QC is not the only solution to the quality problem. The creation of a vin-de-pays category and pulling up vines are being implemented to recover from the exuberant 1980–90s.

Externality Sharing

Winegrowers in all three regions engage in programs to enhance the environmental and the cultural landscape and to fight industrial and urban incursions. There are significant individual costs, for example, installing effluent treatment equipment or maintaining facilities and grounds conducive to patrimonialisation (Gade 2004). These individual costs can be onerous and divisive, such as Napa's laws for protection of rivers (set-backs) and slopes, or controls on winery hospitality functions. Government can step into assist. Bordeaux winegrowers, for example, receive money for effluent control and materials recycling. Generally, however, weaving the territorial fabric depends on the winegrowers' associations. Some costs are shared through organisational dues, but their greater role is mobilisation and political organisation. The NVV's social welfare activities are a good example of the potential and difficulties of organising independent principals to deal with externalities indirectly related to their bottom line. The example is all the more revealing because it is an externality obscured by state subsidy in Bordeaux and Chianti Classico.

Primary workforces are generally protected in more paternalistic large firms, such as the previously family-owned Mondavi (before taken over by Constellation). Many wineries use the health, insurance, and workman's compensation packages offered by the Farm Bureau, the Family Winemakers, or California Association of Wine Grape Growers to offer these programs to full time salaried employees. However, these benefits are becoming difficult to offer due to escalating health and workman's compensation costs (Thach 2002). The casual workforces of the vineyards and hospitality services receive very little in the way of benefits.

To deal with social reproduction, the NVV raised $60 million (to 2006) from its auction for health care, housing and youth programs. Services are delivered indirectly to their workforce via community institutions like the NVV Community Health Center, or directly such as a winery at Napa Valley College or housing for harvest workers. Worker accommodation became an externality when facilities disappeared due to winery expansion, tourist accommodation and a disinclination to have these guests in the neighborhood. Realising the extent of the problem and the lack of individual intention to solve it, the NVV asked the State to impose a tax of $7.76 an acre on vineyard owners to fund housing. The County was asked to allow parcels of land smaller than the agricultural preserve's 40 acre minimum to be used. Vineyard owners approved the tax in a referendum, but construction was only possible when Joseph Phelps donated six acres. Land for five similar facilities is necessary.

Conclusion

Collective organisation enables all three regions to achieve many of the functions of a large integrated corporation: reputation building, marketing, and quality control, even R&D. It has to achieve this cohesion without the authority of ownership or unified brand management, because ownership is fragmented among hundreds or thousands of winegrowers interested in developing their own brands. The comparison of the three regions reveals that many economies of scale and scope are universal needs. Even the lack of managed quality control in Napa is compensated for by its attractiveness to competitive capital. Similarly, the lack of social welfare efforts by Bordeaux and Chianti Classico results from the state providing these services – yet, all three collectives make great efforts to enhance their cultural and environmental landscapes.

In providing these functions the regions make significant efforts to accommodate differentiation, either formally or informally. Examples of the former include obtaining transparency between buyers and sellers, appellations, and classification systems. The recognition of *terroir* as a basis for differentiation is the most pervasive informal accommodation of differentiation, and Bordeaux's futures market exemplifies how the market can evolve to accommodate differentiation. Of course, the informal and informal are interrelated. The winegrower associations in all regions support *terroir* with research and publications extolling their landscape's diversity and several institutions support the operation of the *en primeur* sales.

The regions respond to collectivity and differentiation needs with mechanisms appropriate or driven by their regulatory and sectoral regimes, but the tensions between collectivity and differentiation remains in evolution. Napa obtained *sui generis* laws to protect its reputation when these were lacking in the US legal system, and laws to maintain the dominance of Napa over sub-appellations. It is still working to change the restrictions of the distribution system. Chianti Classico, fought for 50 years to attain autonomy over quality control, but must heal divisions developed during the struggle. Some Bordeaux syndicats have exploited the AOC system's flexibility to increase quality expectations and controls, while others lag in raising the level of their bulk producers. The regional reputation suffers as a result. Thus, as within any organisation, centripetal and centrifugal forces are a constant source of tension. It is the self-governance capacities of the winegrowers that determine the viability of their regional projects. Insofar as these three regions remain benchmarks for wine quality and the value of estate wine continues to increase, the institutions of horizontal integration developed thus far are achieving their objectives.

Acknowledgements

The author would like to thank Philipe Roudie, the good folks at CERVIN and the help of the referees.

References

Arfini, F. (2000), 'The value of typical products: the case of Prosciutto di Parma and Parmigiano Reggiano cheese', in Sylvander, B., Barjolle, D. and Arfini F. (eds), *The Socio-Economics of Origin Labelled Products in Agri-Food Supply Chains: Spatial, Institutional, and Co-ordination Aspects*, Versaillesa, INRA Editions, no. 17–1, pp. 77–97.

Barjolle, D. and Sylvander B. (2000), 'Some factors of success for origin labelled products in agri-food supply chains in Europe: market, internal resources and institutions; in Sylvander B., Barjolle D. and Arfini F. (eds), *The Socio-Economics of Origin Labelled Products in Agri-Food Supply Chains: Spatial, Institutional, and Co-ordination Aspects*, Versaillesa, INRA Editions, no. 17–1, pp. 45–71.

Belletti, G., Brunori G., Marescotti A. and Adanella R. (2002), 'Individual and collective levels in multifunctional agriculture', <http://www.gis-syal.agropolis.fr/Syal2002/FR/Atelier%205/BELLETTI%20MARESCOTTI.pdf>.

César, G. (2002), No. 349 Sénat session extraordinaire de 2001–2002, rapport d'information fait au nom de la commission des affaires éconmiques et du plan (1) par le group de travail (2) sur L'avenir de la viticulture française. <http://www.senat.fr/rap/r01-349/r01-349.html>.

Coelho, A.M. and Rastoin, J.L. (2004), 'Globalization du marché du vin et restructuration des enterprises multinationales' (Globalization of the wine industry and the restructuring of multinational enterprises), paper presented at Oenometrie XI Université de Borugogne, Dijon, May 21–2.

Crespi, J. and Marette, S. (2002), 'Generic advertising and product differentiation', *American Journal of Agricultural Economics*, vol. 84, pp. 691–701.

Escudero, S. (2001), 'International Protection of Geographical Indications and Developing Countries', South Centre, Trade Related Agenda Development and Equity, Working Paper 10.

Gade, D. (2004), 'Tradition, territory, and *Terroir* in French viniculture: Cassis, France, and appellation contrôlée', *Annals of the Association of American Geographers*, vol. 94, pp. 848–67.

Giraud-Héraud, E., Soler, L.G. and Tanguy, H. (2000), 'La relation vignoble-négoce: efficacité de structures verticals différenciées? (The vineyard-merchant relation: efficiency of differentiated vertical structures?)' in Sylvander B., Barjolle D. and Arfini F. (eds), *The Socio-Economics of Origin Labelled Products in Agri-Food Supply Chains: Spatial, Institutional, and Co-ordination Aspects*. Versaillesa: INRA Editions, no. 17–2, 163–75.

Giraud-Héraud E., Soler L.G., Steinmetz S. and Tanguy, H. (1998), 'La regulation interprofessionelle das le secteur vitivinicole est-elle fondée economiquement? (Is regulation by the interprofession in the wine industry economically founded?)' Bulletin de L'O.I.V. 813–14, 1060–84.

Landon S. and Smith C.E. (1997), 'The use of quality and reputation indicators by consumers: the case of Bordeaux wine', *Journal of Consumer Policy*, vol. 20, pp. 289–323.

Lewis N., Moran, W., Perrier-Cornet, P. and Barker, J. (2002), 'Territoriality, enterprise and réglementation in industry governance', *Progress in Human Geography*, vol. 26 pp. 433–62.

Marette, S. and Zago, A. (2003), 'Quality and International Trade: What Strategies for EU AOC system', paper presented at International Conference, Agricultural Policy Reform and the WTO: Where are we Heading?', Capri, June 23–6.

Moran, W. (1993), 'The wine appellation as territory in France and California', *Annals of the Association of American Geographers*, vol. 83, pp. 694–717.

Orth, U.R. (2005), 'Creating and managing regional umbrella brands: a comprehensive quantitative approach', Refereed paper presented at International Wine Marketing Symposium, Sonoma State University.

Roudié, P. (1988), *Vignobles et Vignerons du Bordelais* (1850–1980), Bordeaux, Presses Universitares de Bordeaux.

Roudié, P. (1991), 'Les Viticulteurs de Saint – Emilien', in Candau, J., Roudié, P. and Ruffe, C. (eds), *Saint-Emilion: Terroir et Espace de Vie Sociale*, Maison des Sciences de L'Homme D'Aquitane, Bordeaux, pp. 101–29.

Scott, M.F. and Podolny, J.M. (2002), 'Love or money? The effects of owner motivation in the California wine industry', *Journal of Industrial Economics*, pp. 431–56.

Shelton T. (2001), 'Product differentiation', in Moulton, K. and Lapsley, J. (eds), *Successful Wine Marketing*, Aspen Publishers, Gaithersburg, pp. 99–106.

Schamel, G. and Anderson, K. (2003), 'Wine quality and varietal, regional and winery reputations: hedonic prices for Australia and New Zealand', *The Economic Record*, vol. 79, pp. 357–69.

Swaminathan A. (2001), 'Resource partitioning and the evolution of specialist organizations: The role of location and identity in the U.S. wine industry', *Academy of Management Journal*, vol. 44, pp. 1169–86.

Thach, L. (2002), 'Social sustainability in the wine community: managing for employee productivity and satisfaction', *Wine Business Monthly*, 2002, vol. IX, no. 7.

Tirole, J. (1996), 'A theory of collective reputations (with applications to the persistence of corruption and to firm quality)', *The Review of Economic Studies*, vol. 63, pp. 1–22.

Winfree, J. and McCluskey, J. (2005), 'Collective reputation and quality', *American Journal of Agricultural Economics*, vol. 87, pp. 206–13.

Wrigley, N. (2001), 'The consolidation wave in U.S. food retailing: A European perspective', *Agribusiness*, vol. 17, pp. 489.

Chapter 14

Creating Trust Through Branding: The Case of Northwest Ohio's Greenhouse Cluster

Neil Reid and Michael C. Carroll

Introduction

Northwest Ohio is one of North America's leading regions for the production of greenhouse crops[1] (Reid and Carroll 2005, Figure 14.1). In recent years, however, northwest Ohio's greenhouse industry has faced intense competitive pressure from greenhouses located just north of the U.S.-Canadian border in southern Ontario. Further exacerbating the competitive pressures facing the industry is the fact that the northwest Ohio region has some of the highest energy costs in North America (LaFary et al. 2006; Reid and Carroll 2006). These two factors are creating a very uncertain economic future for an increasingly vulnerable industry.

To address the economic challenges facing their industry a number of greenhouse owners in northwest Ohio decided, at the suggestion of university researchers,[2] to form an industrial cluster. The cluster objective is to bring growers together to identify and collaborate on solving industry-wide challenges that individual greenhouse operations are incapable of addressing as independent business entities. One of the first challenges identified by the growers was their lack of marketing skill. In particular, the industry had no unique 'brand' for their products or region.

The purpose of this chapter is to describe the branding and marketing initiative undertaken by northwest Ohio's greenhouse industry. In particular, we examine the role that this initiative played in increasing levels of social capital among growers. This chapter is divided into six sections. In the next section, section two, we define cluster-based economic development and provide a discussion of collective efficiency. We argue that collective efficiency is the critical ingredient of successful cluster-based economic development initiatives. In section three, we outline the beginning stages of the northwest Ohio greenhouse cluster, emphasising the importance of giving the growers control of the cluster-building process at the earliest possible opportunity. In section four, we describe the decision and criteria used in choosing branding and marketing as the first cluster project. In section five, we describe the development of

---

1 Lucas County, Ohio, ranks in the top five percent of counties in the United States in terms of the value of production of greenhouse crops.

2 The authors.

Figure 14.1   Northwest Ohio

the brand, Maumee Valley Growers, and the evolution of the marketing campaign
that evolved from the brand. We draw some conclusion from our work on this project
in the final section.[3]

Cluster-Based Economic Development

Cluster-based economic development has received considerable attention in the
academic literature. It is an approach that, not surprisingly, has both its adherents
(Porter 1998; Lundequist and Power 2002) and detractors (Martin and Sunley 2003;

_____

3 We stress that this is an ongoing, multi-year, project. Reported in this chapter is the
status of the project at the time of writing (May 2006).

Taylor 2006). Our purpose here is to describe our activities with respect to helping the greenhouse industry in northwest Ohio become more competitive while recognising the controversy surrounding cluster-based economic development.

A cluster is a geographic concentration of businesses in a particular industry. Paraphrasing Porter (1998), a cluster includes all the members of an industry's supply chain who are geographically proximate. It also includes any other entities, such as universities and economic development agencies that can bring value to the industry. Clusters come to life when the members engage in collaborative problem identification and collaborative problem solving. The nature of collaboration varies from cluster to cluster and reflects the unique competitive challenges facing a particular industry in a particular location at particular point in time. Collaboration enhances the competitive economic position of individual businesses and the region in which they are located. In our opinion, collaborating to solve unsolved problems is the essence of cluster-based economic development.

As we have argued elsewhere (Reid and Carroll 2006) the critical ingredient of any successful cluster-based economic development initiative is the ability to harness the power of what Schmitz (1995) terms collective efficiency. Collective efficiency is defined as 'the competitive advantage derived from external economies and joint action' (Schmitz and Nadvi 1999, Figure 14.2). External economies of scale is an old idea according to the economic geography literature and appear as a basic concept in any textbook on the topic (e.g. see Dicken and Lloyd 1990). External economies of scale refer to the business advantages that result from being in close geographic proximity to other businesses in the same industry. Examples of such economies include the availability of a specialised labour pool, access to specialised service providers, and the opportunity to share the costs of specialised transportation infrastructure. External economies of scale are passive in the sense that producers need do little more than locate in a particular geographic area in order to take have access to and take advantage of them. Joint action, on the other hand, is not passive. Joint action only happens when producers consciously decide to collaborate on a particular initiative or set of initiatives.

Joint action is often hindered by a low level of social capital within an industry in a particular geographic area. Rediscovered by social scientists in the 1950s (Putnam 2000) the concept of social capital has its genesis in the work of Lyda Hanifan (Putnam 2000). Social capital is defined by Cohen and Prusak (2001, 4) as 'the stock of active connections among people: the trust, mutual understanding, and shared values and behaviors that bind the members of human networks and communities *and make cooperative action possible*' (emphasis added). The stock of social capital within a community is dependent upon both the quantity and the quality of interactions between its members. In building social capital within a community it is, therefore, necessary to create sufficient opportunities for positive interaction to occur (Falk and Kilpatrick 2000). Building up levels of social capital, where they are below critical threshold levels, is critical for effective joint action to occur (Figure 14.2).

A major challenge to harnessing the potential benefits of joint action is to identify a mechanism to increase levels of trust, mutual understanding, and shared values between competitors. In their highly informative research on the emergence of the Danish co-operative dairy movement of the late-19th century, Svendsen and Svendsen

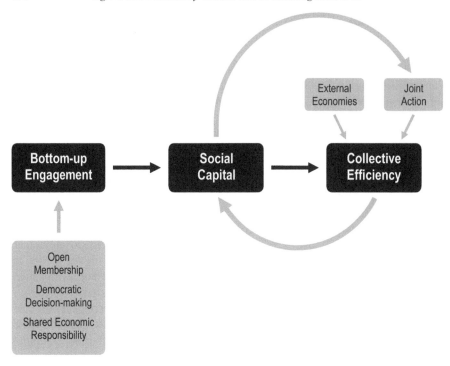

**Figure 14.2   The relationship between bottom-up engagement, social capital
and collective efficiency**

(2000) identify three key characteristics of successful joint action initiatives: open
membership, democratic decision-making, and shared economic responsibility. A
democratic organisational culture emphasising bottom-up engagement and the free
flow of ideas can provide a rich framework for innovative thinking and the building-
up of levels of social capital (Ind 2003). However, for these three characteristics to
become meaningful and to come into play it is necessary that a common economic
interest exist (Figure 14.2).

Development of a common economic interest, which provides the catalyst for
joint action, is often triggered by an exogenous crisis which threatens the economic
viability of an industry. This threat may come from a variety of different sources.
For example, the threat may result from a government reducing the level of
protectionism for domestic producers (Meyer-Stamer 1998; Knorringa 1999; Tewari
1999), intensified competition from foreign producers (Schmitz 1995; 1999), higher
product quality standards imposed by foreign governments (Nadvi 1999), and the
need to meet government-imposed environmental standards (Kennedy 1999). In
the case of the Danish livestock and dairy industry the exogenous threat appeared
in 1879 when the German government imposed a heavy import duty on Danish
livestock. Regardless of the nature of the threat, joint action has allowed different
industries in different regions of the world at different times to successfully respond
to competitive threats.

Responding to an exogenous threat requires thinking differently about the competitive environment in which an industry operates. It requires a willingness to understand the nature of the competitive threat and to experiment with new and innovative competitive strategies. The very process by which an industry learns about its competitive challenges, and moves through the identification of solutions, builds social capital. Learning can, and should, be an interactive process that contributes to change. And if properly managed the learning process can 'result in a capacity to take on new values and attitudes' (Falk and Kilpatrick 2000, 91).

Effective joint action requires entrepreneurial leadership. In the case of the Danish dairy industry the formation (*cytogenesis*) and geographic diffusion (*mitosis*) of cooperatives throughout the country depended upon regional entrepreneurial leaders who were recruited to spread the gospel of cooperative dairying to all parts of the country. These entrepreneurs were trustworthy, experienced, and were capable of inspiring confidence among their peers (Svendsen and Svendsen 2000).

Having identified what we believe to be the key ingredients of a successful cluster-based initiative we now move to a discussion of our attempts to foster social capital within northwest Ohio's greenhouse industry. We begin with a description of the exogenous crises that provided the catalyst for the region's greenhouse growers to begin the process of engaging in joint action initiatives.

Creating the Cluster: A Brief History

In August of 2003 we received funding from the U.S. Department of Agriculture to assess the competitive position of the greenhouse industry in northwest Ohio. As part of the assessment process, we conducted a survey of all greenhouse owners in the five-county region (Figure 14.1).[4] The survey data were supplemented by secondary data, government reports on the status of the North American greenhouse industry, and interviews of local Agricultural Research Service Agents[5] and greenhouse owners. This allowed us to identify the strengths and challenges facing the industry.

In October 2004 we presented our findings to eight of the region's growers, chosen with the assistance of local representatives of the Agricultural Research Service. These growers were considered among the most forward-thinking in the region. They also represented both the retail and wholesale segments of the industry and included what Falk and Kilpatrick (2000, 95) refer to as the 'movers and shakers' and the 'quiet achievers'. This is consistent with the philosophy adopted by the Danish cooperative dairy movement which recognised the important roles played by entrepreneurial individuals in the success of that co-operative effort (Svendsen and Svendsen 2000). Our presentation painted a picture of an industry that is technologically unsophisticated, has an ageing infrastructure, lacks a significant

4 A mail survey of northwest Ohio greenhouse growers was conducted during the period, March–May 2004. The survey was conducted using the Dillman (1978) method. The survey was sent to all 82 growers in northwest Ohio. Twenty-seven surveys were returned, yielding a response rate of 33 percent.

5 The Agricultural Research Service is the chief scientific research agency of the U.S. Department of Agriculture (www.ars.usda.gov).

market presence, relies on traditional sources of fuel, and has little access to capital. Global competition, price wars, Big Box store purchasing agreements, and high and rising utility costs were presented as threats to the future economically viability of the industry. We also highlighted the increasing trade gap between Canada and the United States and between Ohio and Ontario in floriculture products.[6]

We also presented the growers with our assessment of the opportunities awaiting the industry, despite the numerous and obvious challenges facing them. In particular, we stressed that if competing growers were willing to collaborate with each other, then they may be able to successfully address some of the endemic, industry-wide, problems that had plagued them for years. Finally, we presented them with the possibilities of their own inaction. Failure to act in a decisive fashion could result in the demise of a large segment of northwest Ohio's greenhouse industry.[7] Ultimately, however, the challenge was to maintain the existing scale (small) and culture (independent family-owned) of the industry while increasing its competitiveness. As noted by Lee *et al.* (2005, 276), we were entering into 'a process of negotiation between maintaining valued aspects of society, economy, and environment, and engendering new approaches to them'.

If the growers were willing to enter into this process, we promised them that they would be given the opportunity to take ownership of the process and that the evolution of the cluster in terms of vision and management would reflect their hopes and aspirations. Eventually, our role would be relegated to simply helping where we could and being a resource for the cluster. Despite our envisioned limited management role, we took on the responsibility of obtaining financial support for the cluster, until such times as it became financially self-sustaining.

The discussion following our presentation can only be described as tense. The growers in attendance were suspicious. They had been promised solutions to their problems by both government and academia before, only to see these efforts fail to deliver. Their level of trust in us 'the academics' was very low. One grower commented that had it not been raining outside they probably would not have been in attendance. However, the monies to conduct the survey had been secured by a U.S. Congresswoman whom the growers had a good relationship with. She was interested in helping the greenhouse industry in northwest Ohio and, over the years, had gotten to know many of the growers personally. Also, the fact that the Congresswoman's Special Assistant was in attendance at the meeting and was supportive of our efforts enhanced our credibility with the growers. By the time of the presentation, we had also secured a second year of funding from the USDA to pursue a cluster-based strategy for the greenhouse industry in northwest Ohio. Therefore, we were able to bring a significant amount of money to the table to support the implementation of a cluster-based strategy. The growers agreed to spend several weeks digesting the

---

6 Between 1995 and 2004 the size of the trade gap in floriculture products between Ontario and Ohio had increased from $531,186 to $2,014,171, in favor of the Canadians (Industry Canada 2005).

7 Our survey of northwest Ohio growers showed that 40 percent of respondents believed that their industry would be less profitable within the next five years, while 15 percent were planning to downsize or close.

information presented at the meeting and assented to a follow-up meeting in several months to further discuss the prospects of forming a cluster.

In December 2004 at the follow-up meeting the same eight growers agreed to pursue a cluster-based strategy for their industry. One of the growers in attendance noted that 'the region's greenhouse industry is having a tough time. If we don't pull together to identify and solve problems we will surely lose a number of greenhouses' (Center for Regional Development 2006). It was agreed that the concept should be presented to the wider body of the region's growers. The winter conference of the Toledo Area Flower and Vegetable Growers Association in January 2005 provided the perfect opportunity to do this (Carroll and Reid 2005).

Having introduced the concept of cluster-based economic development to the wider population of growers in the region, the next step was to put in place the necessary infrastructure to operationalise the cluster. An Advisory Board was established (January 2005) and a Project Manager and a Champion hired (January 2005 and May 2005 respectively).

The roles of the Advisory Board, Project Manager, and Champion are described in detail elsewhere (Reid and Carroll 2006). With regard to this discussion, however, it is important to reiterate the structure and role of the Advisory Board. The 14 member Advisory Board comprises eight growers and six individuals from academia, government, and economic development. The numerical dominance of the growers on the Advisory Board is intentional and is consistent with the philosophy of bottom-up engagement and of the growers having significant amounts of control and ownership of the process. Grower control is further emphasised by the fact that the growers are the only voting members of the Advisory Board. To extend the democracy of the cluster even further, a decision was made very early on that even growers who were not members of the Advisory Board, could vote on issues if they were in attendance at the meeting at which a particular issue was voted upon.

Branding: The First Project

With the cluster infrastructure now in place, it was time to move on to the identification and implementation of the first project. This was a critical decision. The first project had to meet four criteria. First, it had to require an attainable level of collaboration among the growers. For example, there was little likelihood of the growers being willing to share sensitive information at this early stage of the initiative. Likewise, a project that required the growers to invest significant amounts of their own resources was unlikely to have much chance of getting off the ground. Second, it had to meet the needs of the growers and bring demonstrated value to the industry. Third, it had to provide an opportunity to increase the stock of social capital among the growers. Fourth, it had to have a high probability of success. Failure at this early stage could stall the cluster initiative before it had progressed beyond its infancy. If all of these criteria could be met, then the growers would, hopefully, see the benefits of collaboration and be willing to invest in ongoing collaborative efforts.

With these criteria in mind the Advisory Board sat down and discussed a variety of possible first projects. Two immediately floated to the top of the list – marketing and energy costs, both pressing needs of the industry. The growers admitted that their marketing efforts were fragmented and naïve in their execution. The industry was also at a significant competitive disadvantage because of endemically high energy costs (McKinnon 2006). As the discussion ensued it became clear that marketing of the industry was the best choice for the first project. Because of the relative probability of success, the growers felt that it would be easier to make significant progress on the development of a coordinated marketing effort than on an effort to reduce energy costs. Development of a marketing strategy also met an identifiable business need for the growers. Sixty-five percent of respondents of our 2004 survey had identified a lack of marketing expertise as a significant barrier to expansion.

Almost anything can be branded. This includes companies (Melewar 2005), industries (Hall 2005), places (David 2004; Winfield-Pfefferkorn 2005), and people (DiCarlo 2004). For customers, a strong brand represents the promise of a consistently high quality product or service (Dace 2004). Also, brands often streamline a consumer's decision-making process. For example, when consumers have limited knowledge of a product line, familiarity with a brand name may assist them in arriving at a product choice (Hoeffler and Keller 2003). For producers, a strong brand provides customer loyalty, higher and more predictable sales and profits, premium pricing opportunities, greater return on marketing dollars invested, and easier market entry for new products or services carrying the brand name (Dace 2004; Hoeffler and Keller 2003; Sullivan 1998). Also, brands (unlike technology or products) are extremely difficult to copy and replicate due to their nebulous nature (Dace 2004). They, therefore, represent a wise investment of resources with a potentially large return on investment.

Developing the correct brand identity for a product or service is critical. Once established in the mind of the consumer it is extremely difficult to change consumer perception of a brand. Therefore decisions made early on in the brand development process become critical (Hoeffler and Keller 2003). The key to long-term brand success is the customer experience, and the real measure of brand success is the volume of repeat customers that a brand is able to generate (Ind 2003).

Critical to brand success are the attitudes and behavior of employees. Employee attitudes and behaviors directly impact the customer experience. If employees are to provide customers with a positive consumer experience, it is necessary that they themselves believe in the brand, the organisation, and products that the brand represents. As noted by De Chernatony (2001, 32), 'staff have unique knowledge and skills which enable them to deliver the brand's functional claims, and when their personal values align with those of the brand, they are committed emotionally to delivering the brand'. Employee attitudes are, in turn, influenced by the corporate culture of the organisation whose products or services are being branded. Organisational culture, therefore, becomes a critical ingredient of brand success (De Chernatony 2001).

Maumee Valley Growers

The Advisory Board recognised the need to engage the assistance of a professional branding and marketing firm, and in keeping with the cluster philosophy of utilising local expertise to solve local industry problems, they chose a local firm to develop a comprehensive marketing strategy. The agency chosen by the growers has many years of experience in serving a wide variety of clients ranging from local police departments to multinational automotive companies. The agency was officially engaged by the Advisory Board in August 2005.

One of the first tasks undertaken by the agency was to attain a greater understanding of the industry. They attended Advisory Board meetings and, more importantly, spent time in the field with the Champion as he visited growers. These visits by an agency representative also provided opportunities to make growers aware of the upcoming branding initiative, to educate the growers on the concept of branding, and to answer any questions that the growers might have regarding the concept.

By mid-September 2005, the process of developing a brand name, logo, and positioning statement for the northwest Ohio greenhouse industry was undertaken. As the brand development moved forward the Advisory Board remained highly engaged. Particularly, growers were asked to provide suggestions for both the brand name and positioning statements. Through an iterative process that included market

Figure 14.3   Brand name, logo, and positioning statement for the Northwest Ohio greenhouse cluster

testing of brand names, logos, and positioning statements the Advisory Board were presented with the finalists in each of these categories. In keeping with the tradition of a grower driven decision-making process, only the growers in attendance at the meeting were allowed to vote on the final brand choices (Figure 14.3).

The brand name chosen was *Maumee Valley Growers*. The name reflects the geography of the region in which the northwest Ohio greenhouse industry is located.[8] The positioning statement, *Choose the Very Best*, conveys the message that purchasing locally-grown plants ensures that the customer is purchasing the highest quality product available in the region.

---

8 The appellation, Maumee Valley, refers to the valley of the Maumee River that starts in Fort Wayne, Indiana and flows through northwest Ohio before draining into nearby Lake Erie.

With the brand established, the Maumee Valley Growers (MVG) has started the process of capitalising on the brand through the development and implementation of a comprehensive marketing campaign. Articles and advertisements are appearing in the local newspaper and on a local television station to coincide with the start of the spring planting season (McKinnon 2005; Toledo Business Journal 2006). A new website has been established (www.maumeevalleygrowers.com). This website is aimed at both customers and greenhouse owners. For example, customers can use the website to find a greenhouse close to their home. It is also used to communicate with greenhouse owners regarding activities and opportunities associated with the cluster. The website is also being utilised to increase grower engagement in the project. Local greenhouse owners can provide gardening tips for consumers while at the same time advertising their own greenhouse.

Conclusion

The process of branding the northwest Ohio's greenhouse industry is as much about building up levels of social capital and moving towards the creation of a collective identity (Fromhold-Eisebith 2006) as it is about marketing the industry to local consumers. It is about raising the self-confidence of each grower, with the aim that they will realise that they have the capacity to bring about positive change. In turn, this will bring them a more secure economic future (Ray 1998). If the branding process is to achieve these objectives it must engage the growers. As David states (2004), the brand, and the process by which it is created, must 'come from the hearts' of the growers.

A major question hanging over the branding efforts of the northwest Ohio greenhouse growers focuses upon the extent to which this effort has caused a significant and long-lasting change in terms of willingness to think and act collaboratively. Or will the brand simply become 'clothing' that is put on for outsiders (Lee *et al.* 2005)? To what extent will the overarching identity of the brand be able to successfully embrace the large number of 'mobile identities' (Lee *et al.* 2005, 275) that currently exist within the northwest Ohio greenhouse industry? If the branding effort is successful, to what extent can the process of building the brand elevate the level of social capital within the industry so that other collaborative efforts might be attempted (Lee *et al.* 2005)?

Acknowledgements

This research is funded by U.S. Department of Agriculture grants CSREES 2003-06230, CSREES 2004-06222, and CSREES 2005-02216. The research team comprises faculty and staff from Bowling Green State University, Indiana State University, The Ohio State University, The University of Toledo, and Toledo Botanical Gardens.

References

'Dictionary of Marketing Terms', American Marketing Association, <www. marketingpower.com>.

Carroll, M. and Reid, N. (2005), 'Competing successfully in a global marketplace', Toledo Area Winter Greenhouse Conference, Monclova, Ohio, 13 January.

Center for Regional Development Bowling Green State University (2006), 'Growing the greenhouse industry' *Centerlines*, Winter, vol. 2.

Cohen, D. and Prusak, L. (2001), *In Good Company, How Social Capital Makes Organizations Work*, Harvard Business School, Boston.

Dace, C. (2004), 'What's in a name?', *IEE Engineering Management*, February/ March, pp. 38–41.

David, S. (2004), 'Looking at life in a different light: The branding of Australia', *Screen Education*, vol. 36, pp. 16–20.

De Chernatony, L. (2001), 'A model for strategically building brands', *Journal of Brand Management*, vol. 9, no. 1, pp. 32–44.

DiCarlo, L. (2004), 'Six degrees of Tiger Woods', <www.forbes.com>, accessed 8 May 2006.

Dicken, P. and Lloyd, P.E. (1990), *Location in Space* (3rd ed.), Harper & Row, New York.

Falk, I. and Kilpatrick, S. (2000), 'What *is* social capital? A study of interaction in a rural community', *Sociologia Ruralis*, vol. 40, no.1, pp. 87–110.

Fromhold-Eisebith, M. (2006), 'Which mode of (CLUSTER) promotion for which aspect of entrepreneurship? A differentiating view on institutional support of automotive clusters', in Gatrell, J.D. and Reid, N. (eds), *Enterprising Worlds: A Geographic Perspective on Economics, Environments, and Ethics*, Springer, Dordrecht, pp. 13–28.

Hall, J. (2005), 'Re-establishing trust in financial services', *Brand Management*, vol. 12, no. 4, pp. 232–5.

Hanifan, L.J. (1916), 'The rural school community center', *Annals of the American Academy of Political and Social Science*, vol. 67, pp. 130–8.

Hoeffler, S. and Keller, K.L. (2003), 'The marketing advantages of strong brands', *Journal of Brand Management*, vol. 10, no. 6, pp. 421–45.

Ind, N. (2003), 'Inside out: how employees build value', *Journal of Brand Management*, vol. 10, no. 6, pp. 393–402.

Industry Canada (2005), *Strategis: Canada's Business and Consumer Site*, <http://www.strategis.gc.ca>, accessed 8 May 2006.

Kennedy, L. (1999), 'Cooperating for survival: tannery pollution and joint action in the Palar Valley (India)', *World Development*, vol. 27, no. 9, pp. 1673–91.

Knorringa, P. (1999), 'Agra: An old cluster facing new competition', *World Development*, vol. 27, no. 9, pp. 1587–1604.

LaFary, E., Gatrell, J.D., Reid, N. and Lindquist, P.S. (2006), 'Specialized agriculture: local markets and global competitors in Ohio's greenhouse industry', in Gatrell, J.D. and Reid, N. (eds), *Enterprising Worlds: A Geographic Perspective on Economics, Environments, and Ethics*, Springer: Dordrecht, pp. 57–70.

Lee, J., Arnason, A., Nightingale, A. and Shucksmith, M. (2005), 'Networking: social capital and identities in European rural development', *Sociologia Ruralis*, vol. 45, no. 4, pp. 269–83.

Lundequist, P. and Power, D. (2002), 'Putting Porter into practice? Practices of regional cluster building: Evidence from Sweden', *European Planning Studies*, vol. 10, no. 6, pp. 685–704.

Martin, R. and Sunley, P. (2003), 'Deconstructing clusters: chaotic concept or policy panacea', *Journal of Economic Geography*, vol. 10, no. 6, pp. 685–704.

McKinnon, J. (2005), 'Greenhouses in area tend to marketing', *Toledo Blade*, 8 December, <www.toledoblade.com>, accessed 8 May 2006.

McKinnon, J. (2006), 'Greenhouses likely to up prices: Area growers hit by high fuel costs', *Toledo Blade*, 28 March, <www.toledoblade.com>, accessed 8 May 2006.

Melewar, T.C., Hussey, G. and Srivoravilai, N. (2005), 'Corporate visual identity: the re-branding of France Telecom', *Journal of Brand Management*, vol. 12, no. 5, pp. 379–94.

Meyer-Stamer, J. (1998), 'Path dependence in regional development: persistence and change in three industrial clusters in Santa Catarina, Brazil', *World Development*, vol. 26, no. 8, pp. 1495–511.

Nadvi, K. (1999), 'Collective efficiency and collective failure: The response of the Sialkot surgical instrument cluster to global quality pressures', *World Development*, vol. 27, no. 9, pp. 1605–16.

Porter, M.E. (1998), 'Clusters and the new economics of competition', *Harvard Business Review*, November–December, pp. 77–90.

Putnam, R.D. (2000), *Bowling Alone*, Simon & Schuster, New York.

Ray, C. (1998), 'Culture, intellectual property and territorial rural development', *Sociologia Ruralis*, vol. 38, no. 1, pp. 3–20.

Reid, N. and Carroll, M.C. (2005), 'Using cluster-based economic development to enhance the economic competitiveness of northwest Ohio's greenhouse nursery industry', *Papers of the Applied Geography Conferences*, vol. 28, pp. 309–19.

Reid, N. and Carroll, M.C. (2006), 'Collaborating to compete: The case of northwest Ohio's greenhouse industry', in Gatrell, J.D. and Reid, N. (eds), *Enterprising Worlds: A Geographic Perspective on Economics, Environments, and Ethics*, Springer: Dordrecht, pp. 41–56.

Schmitz, H. (1995), 'Collective efficiency: Growth path for small-scale industry.' *The Journal of Development Studies*, vol. 31, no.4, pp. 529–66.

Schmitz, H. (1995), 'Small shoemakers and fordist giants: tale of a supercluster', *World Development*, vol. 23, no. 1, pp. 9–28.

Schmitz, H. (1999), 'Global competition and local cooperation: success and failure in the Sinos Valley, Brazil', *World Development*, vol. 27, no. 9, pp. 1627–50.

Schmitz, H. and Nadvi, K. (1999), 'Clustering and industrialization: introduction', *World Development*, vol. 27, no. 9, pp. 1503–14.

Sullivan, M.W. (1998), 'How brand names affect the demand for twin automobiles', *Journal of Marketing Research*, vol. 35, no. 2, pp. 154–65.

Svendsen, G.L.H. and Svendsen G.T. (2000), 'Measuring social capital: the Danish co-operative dairy movement', *Sociologia Ruralis*, vol. 40, no. 1, pp. 72–86.

Taylor, M. (2006), 'Clusters and local economic growth: unpacking the cluster model', in Gatrell J.D. and Reid, N. (eds), *Enterprising Worlds: A Geographic Perspective on Economics, Environments, and Ethics*, Springer: Dordrecht, pp. 100–17.

Tewari, M. (1999), 'Successful adjustment in Indian industry: the case of Ludhiana's woolen knitwear cluster', *World Development*, vol. 29, no. 9, pp. 1651–71.

Toledo Business Journal (2006), 'Greenhouse cluster targets brand awareness', *Toledo Business Journal*, March, vol. 1, pp. 22, 28.

Winfield-Pfefferkorn, J. (2005), The Branding of Cities, unpublished M.A. Thesis, Syracuse University.

Chapter 15

Growing a Global Resource-Based Company from New Zealand: The Case of Dairy Giant Fonterra[1]

Christina Stringer, Christine Tamásy, Richard Le Heron
and Stuart Gray

Introduction

'Does geography matter?' is a question that particularly fascinates academics based
in New Zealand, a market-oriented island economy located in the south west Pacific.
We would like to answer this question with an emphatic 'yes' and suggest that more
attention be given to geographic connections being created in the global economy.
Using Fonterra Cooperative Ltd, the New Zealand based-dairy giant ranked sixth
in the world in terms of revenue, as a case study we discuss how we came to this
conclusion. Fonterra illustrates that it is indeed possible to grow a global company
from 'the edge' of the global economy and raises doubts about arguments pertaining
to disadvantages of remoteness and is a particularly interesting example of an actor
on the resource periphery (Hayter *et al.* 2003).

Fonterra was formed in 2001 through the merger of the New Zealand Dairy
Board (NZDB), the export marketing arm of the New Zealand dairy industry,
and two large dairy co-operative companies – the New Zealand Dairy Group and
Kiwi Co-operative Dairies. Fonterra, New Zealand's largest company is owned by
11,200 New Zealand dairy farmers; 95 percent of all New Zealand dairy farmers are
supplying shareholders of the co-operative. Fonterra represents today more than 20
percent of the country's export earnings and 7 percent of GDP. Fonterra was created
as a culmination of 30 years of dairy company rationalisation and mergers within
New Zealand and intense industry debate about the structure needed to drive scale
economies at the supply end in order to counteract market forces with the power to
drive prices down. The dairy industry globally is a localised industry with the vast
majority of the world dairy production consumed in the country of manufacture
predominately as liquid milk. Only 7 percent of product is traded across borders

---

1 This chapter is an intellectual spin-off from the one-day field trip associated with the
IGU Dynamics of Economic Spaces 2006 Conference in Auckland, New Zealand. Delegates
visited the multinational dairy company Fonterra which offers an insightful case study into
a global company headquartered in a location that seems to be disadvantaged by spatial
remoteness.

today and it is in this traded space that Fonterra mainly operates. The New Zealand dairy industry is unusual in that from its foundation it has always had a focus beyond its limited home market. Fonterra, today, is a major player in the global cross-border trade in dairy products. The company's supply chain reaches from the farm to consumers in 140 countries with processing occurring on four continents (Gray *et al.* 2007). Fonterra is engaged in a range of supply agreements and joint venture arrangements and partnerships with global food companies such as Nestlé, Kraft and Masterfoods, in addition to supplying food retail chains and the hospitality sector with its own branded products.

Global lead firms such as Fonterra, which are embedded in increasingly complex networks of business relationships, are seen as the primary movers and shapers of the global economy for three reasons (Dicken 2007; Yeung 2006). Firstly, they are able to coordinate and control various processes and transactions within networks, both within and across national borders. Secondly, they are potentially able to take advantage of geographical differences in the distribution of factors of production and in state policies. Thirdly, they are able to switch resources and operations between locations on a global scale. The aim of this chapter is to explore the globalising activities of Fonterra as a case study that illustrates the reality of globalisation processes. The understanding that comes from examining the created and changing geography of Fonterra forces a re-look at overly simplistic and deterministic analyses of remoteness and geographical isolation[2] assumptions that can abound in discussions on globalisation. The chapter is organised as follows. The first section discusses the conceptual considerations relating to the topic and questions deterministic assertions commonly made regarding the impact of distance. The second section describes the regulatory frameworks in which the networks of the world dairy industry are embedded. We question the often cited claim in the globalisation literature that companies now operate in a borderless world where 'the state' is no longer a major player. In the empirical section, Fonterra is conceptualised as networks of internalised and externalised relationships, which exist between dependent and associated firms. This allows us to investigate diversified forms of intra- and inter-firm networks, such as sourcing and strategic alliances, and to show how these linkages reflect major business development strategies and regulatory conditions. The individual firm's perspective from *within* networks enables us to explore in detail the role of Fonterra as constituent part of wider firm networks 'through which emergent power and effects are realized over space' (Hess and Yeung 2006, 1196). The analysis illustrates how Fonterra seeks to overcome trade barriers through strategies that reflect the geo-political market environment for the dairy industry.

Globalisation, Location and Networks

New Zealand, as a resource periphery economy is largely dependent on earnings from agriculture, forestry, and tourism for its wealth, and its location 'on the edge' of the world economy has always been a point for excited discussions on the presumed

---

2 Huskey (2006) discusses the difficulty of defining 'remote regions'.

disadvantages of remoteness. *The Economist*, for example, tellingly pointed out in November 2000:

> As 'the last bus stop on the planet', New Zealand is at a disadvantage compared with other small economies such as Ireland or Finland. A circle with a radius of 2,200 kilometres centred on Wellington encompasses only 3.8m people and a lot of seagulls. A circle of the same size centred on Helsinki would capture well over 300m people.

Using gravity models Evans and Hughes (2003) calculated that New Zealand is the most remote of the Organisation of Economic Cooperation and Development (OECD) countries. Despite the 'distance destroying' capacity, e.g. electronic communication, often claimed to be incorporated in ongoing globalisation processes (Murray 2006), distance effects are still assumed to translate into additional transaction costs and weakening cultural ties that negatively influence economic relationships and, therefore, constrain national economic growth (e.g. Evans and Hughes 2003). According to Poot (2004) other geographical features generate economic disadvantages in addition to pure distance. New Zealand's small population of about 4 million people and its low population density of about 15 people per square kilometre make it difficult for firms to reap economies of scale in the domestic market. The deterministic assertions relating to distance and related disadvantages, such as the viewpoints published by New Zealand economic 'think-tanks' (e.g. New Zealand Treasury (e.g. Skilling 2001; McCann 2003) and The New Zealand Institute (e.g. Skilling and Boven 2006a; 2006b)), in part seems to have confused the reality of what is going on in the 'real' world of global geographies. Distance and scale disadvantages do not necessarily constrain a company's growth, as Fonterra's growth path perfectly illustrates. Today, the New Zealand market makes up little more than 5 percent of Fonterra's revenues. It is a company that is focused predominantly on the global markets beyond its remote and limited home base.

Fonterra can be seen as an example of how the role of economic actors can overcome perceived geographical disadvantages at different times. This chapter shows that it is not the forces of nature or gravity that primarily influence economic growth but the role of economic actors and institutions seeking the expansion of value (Fields 2004).[3] Furthermore, economic actors and economic processes are mediated by institutional arrangements. Thus, to understand a globalising company such as Fonterra, the role of the state (in different territories and across territories) and the nature of the competitive environment become of central importance. The next section elaborates in more detail why the state in various guises – despite globalisation forces – is still a major regulator of worldwide flows of trade, production and finance in the world dairy scene.

Regulatory Barriers within the Dairy Industry

The dairy industry in New Zealand has a long history of government structured intervention at the sales and marketing end. Since the 1920s successive New

---

3 See Huskey (2006) for the 'institutions hypothesis'.

Zealand governments have sought to maintain a 'single-seller' desk (Gray *et al.* 2007). In 1973 when the United Kingdom joined the European Union, the New Zealand dairy industry which to date had privileged access to the British market, was faced with a life-threatening decision – to retreat from any internationalisation strategy and downsize in order to focus solely on the small domestic market or to aggressively move into markets outside of the European Union. The New Zealand Dairy Board (NZDB) (Fonterra's predecessor) chose the latter and undertook an international diversification strategy. In doing so, they encountered an international dairy market which was, and remains, one of the most heavily regulated agricultural markets with a complex network of domestic and international trade barriers (Figure 15.1). Fonterra (and its predecessor the NZDB), initially as a prominent and then dominant player, had little choice but to meet the regulatory constraints head on, with the ability to access markets being of critical importance for survival and growth. The NZDB established marketing subsidiaries, processing facilities and product development offshore; the diversification strategy was in part determined by protectionist barriers. The NZDB entered into markets with lower trade barriers as well as establishing subsidiaries in higher tariff markets with significant growth potential which were not yet open to trade. Tariff barriers and market access shaped the company's diversification strategy.

In developed markets in particular, governments intervene domestically through market access restrictions and subsidies which distort both production and trade. This creates a situation where exporting countries, such as New Zealand, are at a disadvantage as they are not able to capture price premiums. The world's most valuable dairy markets are Europe, North America and Japan, all of which are subject to a high degree of product access quotas and high tariffs. The New Zealand dairy industry faces excessive tariff barriers in these developed markets; for example,

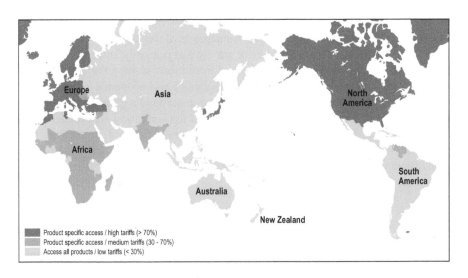

Figure 15.1   World dairy access and tariffs

Source: Fonterra 2006.

tariffs in Japan and Switzerland for some dairy products are as high as 360 percent and 800 percent respectively (Goff 2006). While these percentages may be extremes, even much lower tariff rates still obstruct commodity trade in industries such as dairy where profit margins may be only a few percentage points. Furthermore in accessible markets, multinational competitors benefit from subsidies allowing them to match New Zealand's low-cost position (Cartwright 1990).

The New Zealand government has been actively involved in multi-lateral trade negotiations and pluri-lateral agreements to reduce trade distorting mechanisms; as a low cost exporter New Zealand 'would be able to fully exploit their comparative advantage in undistorted world dairy markets' (Cox and Zhu 2005, 171). During the Uruguay Round, the Agreement on Agriculture (URAA), where agricultural trade was made subject to multilateral trade rules, was introduced. The URAA contained commitments pertaining to the reduction of trade distortions in dairy products including a tariff quota system, a reduction in subsidies and tariffs, and increased access to markets. As an outcome of the Uruguay Round, New Zealand benefited from increased country-specific tariff butter quota access to the European Union, nevertheless, protectionist barriers in major markets remain; developed countries spend in the vicinity of three billion euros annually in subsidies (Waugh 2004).

New Zealand has taken an active role in the Cairns Group which promotes agricultural reform under GATT. Formed in 1986, during the Uruguay Round, in response to increased subsidised exports from the European Union and the United States into non-traditional markets, the Cairns Group today comprises eighteen agricultural exporting countries. Unable to subsidise their own agricultural exports, the countries collectively joined together to advocate agricultural trade reform under the free market system. During rounds of multi-lateral trade negotiations, the Cairns Group's agenda has focused on a substantially more open regime for international agricultural trade. Proposing a non-discriminatory trade regime, the Cairns Group is seeking to have negotiable tariff barriers replace more rigid non-tariff barriers and also to introduce the progressive liberalisation of all tariff barriers and export subsidies. Under the ongoing WTO Doha Development Round, the Global Dairy Alliance, a consortium of six nations all of whom are members of the Cairns Group (Argentina, Australia, Brazil, Chile, New Zealand and Uruguay), is seeking increased market access, elimination of export subsidies and internal domestic support programmes. The suspension of the Doha Round talks in 2006 slowed the progress of liberalisation and was seen as a set-back to the New Zealand dairy industry. 'The proposed reduction in export subsidies would have provided low cost exporters, like New Zealand, with an opportunity to replace high cost, subsidised exports products in world markets' (Rabobank 2006, 5).

New Zealand, Australia and the European Union dominate the international dairy trade. The European Union is New Zealand's main competitor and a principal destination export market. The European Union has agreed to a programme of subsidy reduction in line with WTO requirements and also as a result of internal financial pressures within the European Union to reduce the costs of the Common Agricultural Policy (CAP) (HM Treasury and DEFRA 2005; Kelch and Normile 2004). The reforms are underway with a move from subsidising support for production and export to land stewardship. As a result the European dairy industry is being forced to become more efficient. Rationalisation of smaller dairy companies is underway, for example in the UK, France and Germany while larger companies are actively engaged in reviewing merger opportunities, including potential cross-border mergers (e.g. between Dutch and Danish companies). The European industry appears to be pulling back from commodity export activity to traditional markets such as Africa and the Middle East in order to focus on the higher returning internal European market; the world's largest and most affluent dairy market. At the same time there has been an expansion internationally by some of the larger companies who are more focused on retail brands. The French company Danone is the best example of this. The higher margins achieved by branded retail products such as yoghurts and dairy desserts, coupled with a willingness to invest in local production of these short shelf-life products in countries like China, using locally sourced milk, offers these companies a potential global footprint that is very different to the historic European one of exporting dairy commodities using export subsidies.

Fonterra's Global Network Patterns: Upgrading the Value Chain?

By 2001, the New Zealand dairy industry was overwhelmingly focused on global markets and had built up a suite of expertise which included an international network of sales offices, in-depth market intelligence, and production of quality dairy products which were competitively priced through expertise in scale manufacturing and the ability to scale up new products. One of the biggest constraints Fonterra faced was how to maintain consistent supply to customers in a seasonal industry where demand for product over 12 months could not be easily met from pasture-based New Zealand dairy farmers who produce milk for only 9 months of the year. Internationally competitors were merging to gain efficiencies in production and supply chains. International food companies were seeking players with the ability to supply dairy ingredients when and where they are wanted. If the company was to become the global leader in cross-border dairy trade, it had to be able to source sufficient products to meet customer demand over the full 12 months of the year.

> The past decade has seen unprecedented retail consolidation ... To play in this consolidating market, we need to demonstrate a clear ability to manufacture to forecast, deliver in full, on time and in spec more than 98 percent of the time, offer electronic data interchange facilities and full marketing support and we have to demonstrate that our customers want our product ... Our commodity customers ... no longer look to us to supply ingredients. They want assurances that this supply will be reliable, that it will be at reasonably stable prices, that we will manage geographic risks for them so if drought hits Australia for example, any production shortfall will not impact on them (Ferrier 2005).

In order to work around this critical issue of seasonal milk production, Fonterra is increasingly sourcing milk internationally and has entered into partnerships with foreign dairy companies. These reflect a transition by Fonterra from transactional spot buying of other companies' dairy commodities as required, to a more formalised partnership where Fonterra undertakes to market that company's product internationally in return for ongoing access to that company's milk. One of the challenges the company faces in partnership arrangements is quality control – the need for the quality to be the same as if it was coming out of New Zealand. Third party producers are required to manufacture to Fonterra's specifications, including product packaging and documentation. For example, to drive cost efficiencies and ensure consistent quality control, the documentation for product sourced in Latin America and Australia is handled by the documentation centre in Auckland even though the product itself moves from one third party country to another (e.g. from Latin America to Asia) and is never physically in New Zealand. Through strategic partnerships, Fonterra is gaining market presence in protectionist markets where demand for dairy products has until now been met by local supply. For instance Fonterra may apply its acknowledged expertise in dairy product manufacture to another company's milk. For example in the United States, Fonterra has partnered with the largest United States milk producer, Dairy Farmers of America, to manufacture milk protein products using American milk and New Zealand expertise. Fonterra is now the largest exporter of dairy products from the United States. 'Fonterra is a buyer and an exporter of US non-fat dry milk to other foreign markets, providing valuable global marketing expertise' (Blayney and Gehlhar 2005). The benefits to Fonterra are significant 'selling dairy products from other countries generates significant economies of scale through Fonterra's network of sales offices and mitigates the seasonality of supply from New Zealand' (Gray *et al.* 2007).

In 2007, Fonterra has 56 subsidiary and partnership operations (Figure 15.2). Partnerships can be classified as one of three types – market channel focused, supply focused or innovation-focused. The majority of partnerships are of the market channel type and typically involve linking with an existing company in a market where Fonterra is not yet well established as a leader. However both sourcing and innovation partnerships are of increasing importance and are being established in key markets such as North America, Europe, Latin America and Australia. The concentration of subsidiaries in Asia reflects the historic NZDB focus on this region following the loss of privileged access to the UK market in the 1970s. Asia offered open market access and was the focus of a market development push when the industry was forced to diversify away from the traditional UK market (Gray *et al.* forthcoming). Latin America was also an area of early sales activity. Fonterra has in recent years built on their long established sales operations on that continent and is now actively developing sourcing arrangements with Latin American dairy companies in the so-called Southern Cone. This enables Fonterra to offer an alternative supply source for customers around the world. Fonterra is establishing itself as a major player in this region of low cost production and emerging competition. In turn, these changes in partnership arrangements and increasing complexity of Fonterra's global trade flows have required changes in Fonterra's organisational structure to reflect the evolving corporate geography and the need to match skills with not only its customer base but increasingly also with its supplier base, both of which may be located in any part of the world.

The ability to supply customers globally is a key attraction of Fonterra to large global food companies. In 2002, Fonterra entered into the first of a number of partnerships with global food giant Nestlé S.A., a Swiss-based publicly listed company. Nestlé grew out of an initial base of producing baby formula with no

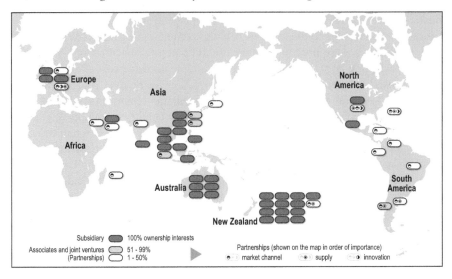

Figure 15.2   Fonterra's global network pattern, 2006

Source: Fonterra 2006.

obvious national Swiss mission (www.nestlé.com). Nestlé is now the world's largest food company with revenues of $US80 billion; as part of this is it also the world's largest dairy company. In contrast, Fonterra is a privately owned farmer co-operative which has evolved from an ongoing national project to grow the New Zealand dairy industry, thus reflecting the critical importance of the industry to the New Zealand national economy. As such, successive New Zealand governments have set the regulatory environment under which Fonterra and its predecessors operated (Gray *et al.* 2007).

In recent years Nestlé has implemented a strategy of divesting from commodity processing and has sold a number of milk processing plants in Latin America and Europe, or entered into a number of joint venture arrangements with Fonterra. In 2003, Dairy Partners America (DPA), a joint venture arrangement between Nestlé and Fonterra, became operational. DPA has established joint ventures in Argentina, Brazil, Colombia, Ecuador, Trinidad and Tobago, and Venezuela. DPA manufactures milk powders in Argentina, Brazil, Columbia and Ecuador and operates chilled dairy facilities in Argentina, Brazil, Ecuador, Trinidad & Tobago, and Venezuela (Fonterra 2004). On a venture-by-venture basis, both companies put in processing facilities and retail business brands into the joint venture operation. Fonterra and Nestlé still retain separate operations in Latin America, for example, Fonterra owns a 57 percent stake in Soprole, the leading dairy company in Chile with the remaining 43 percent owned by the investment arm of the Chilean Catholic Church.

The overlapping geographies of Fonterra and Nestlé are illustrative of partnership developments. Both companies have manufacturing operations and strong consumer brands, with Nestlé stronger in the latter globally. The joint ventures they have entered into are evolving into organisations of shared competencies with Fonterra focusing on the processing of milk and Nestlé on consumer marketing. This pattern of interaction started in the Americas and now includes Australia as Nestlé exits from manufacturing commodity milk powders around the world. Fonterra recently purchased a Nestlé processing plant in Australia which continues to supply products to Nestlé. Overlaying and tying the relationship together is a common interest. Internationally, Nestlé is Fonterra's largest single customer, while Fonterra is the only international dairy supplier large enough to meet Nestlé's volume and geographic requirements. It is a complex relationship; in some markets the two companies remain strong competitors and vigorously compete with each other on the retail shelf (e.g. Asia) while in other markets (e.g. much of Latin America) the two companies work together in co-operation.

Discussion and Conclusion

By focusing on the agency of Fonterra, in the context of the world international dairy environment, we have brought into question a range of claims in the international literature relating to the disadvantages of engaging in the global economy from a peripheral structural and geographical position. Why focus on agency? We argue that since economic and institutional actors shape the expansion of capitalist investment processes we need to avoid being trapped into arriving at pre-determined conclusions. For instance, gravity model calculations produce a gradient that positions location relative to one another. Fonterra's story is doubly valuable. Not only does it illustrate that growth is possible from the edge of the global economy (so demolishing conventional predictions), it exposes some of the inner workings of successful international competition. Fonterra's experience as a global corporate illustrates at least three points: firstly, geography is far more complex than simple physical distance; geography is in fact constructed through relationships and interactions amongst actors, secondly, economies of scale relate to intra- and inter-company organisation and should not be narrowly interpreted in terms of domestic markets, and thirdly, governments continue to differentiate the regulatory landscape faced by companies.

This investigation of Fonterra's strategy and alliances highlights the real constraints of institutional barriers such as government imposed tariffs, and the porosity of borders that comes from the effective use of agency. The evidence of the figures shows how in a globalising world the notion of geography remains extremely valuable – providing its content is constructed from analysis that centres on relationships, interactions and possibilities of action. Companies can, and frequently do move around territorial and institutional barriers. To do so, companies change their capacities, capabilities and organisational arrangements. They may do this by themselves or in combination with other actors with whom they link up in various arrangements. The geography of constraint and opportunity they encounter is rarely static for long. An assessment of a company's geographic identity today may be quite inaccurate tomorrow or next year. Fonterra recently announced that the company will move into a new focus of developing local milk behind borders thus opening up the opportunity for Fonterra to become a bigger player in the geographical market space in which it doesn't currently have a strong position (Stringleman 2007; van der Heyden 2007). From wherever economic geographers attempt to investigate a changing world, it is understandings of created geographies that really matter.

References

Blayney, D. and Gehlhar, M. (2005), 'U.S. dairy at a new crossroads in a global setting', *Amber Waves*, November, <www.ers.usda.gov> (retrieved 12 January 2007).

Cartwright, W. (1990), *The Competitive Advantage of the New Zealand Dairy Industry*, The University of Auckland Graduate School of Business, Auckland, New Zealand.

Cox, T. and Zhu, Y. (2005), 'Dairy: assessing world markets and policy reforms: implications for developing markets', in Aksoy, M.A. and Beghin, J.C. (eds), *Global Agricultural Trade and Developing Countries*, World Bank, Washington D.C.

Dicken, P. (2007) (5th Edn.), *Global Shift: Mapping the Changing Contours of the World Economy*, Sage, London.

Evans, L. and Hughes, P. (2003), Competition Policy in Small Distant Open Economies: Some Lessons from the Economics Literature, Working Paper, 03/31, New Zealand Treasury, Wellington, New Zealand.

Fields, G. (2004), *Territories of Profit*, Stanford University Press, Stanford, USA.

Ferrier, A. (2005), 'Globalisation in dairy – irresistible force or immovable object?' Address to Dairy Industry Newsletter Conference, London, 27 April 2005.

Fonterra Co-operative Limited (2004), Dairy Partners Americas to Operate in Ecuador, Colombia and Trinidad & Tobago <www.fonterra.com> (retrieved 12 January 2007).

Fonterra Co-operative Limited (2006), *Fonterra Annual Report 2006*, Fonterra, Auckland, New Zealand.

Goff, P. (2006), Trade and the NZ Dairy Industry, <www.beehive.govt.nz> (retrieved 15 November 2006).

Gray, S., Le Heron, R., Stringer, C. and Tamásy, C. (2007), 'Competing from the edge of the global economy: the globalising world dairy industry and the emergence of Fonterra's strategic networks', *Die Erde*, vol. 138(2): 127–147.

Hayter, R., Barnes, T. and Bradshaw, M. (2003), 'Relocating resource peripheries to the core of Economic Geography's theorizing: rationale and agenda', *Area*, vol. 35, pp. 15–23.

Hess, M. and Yeung, H.W.-C. (2006), 'Whither global production networks in economic geography? Past, present, and future', *Environment and Planning A*, vol. 38, 1193–204.

HM Treasury and DEFRA (2005), A Vision for the Common Agricultural Policy <www.defra-gov.uk> (retrieved 23 January 2007).

Huskey, L. (2006), 'Limits to growth: remote regions, remote institutions', *The Annals of Regional Science*, vol. 40, 147–55.

Kelch, D. and Normile, M.A. (2004), CAP reform of 2003–04, *United States Department of Agriculture (USDA)* <www.ers.usda.gov> (retrieved 23 January 2007).

McCann, P. (2003), Geography, Trade and Growth: Problems and Possibilities for the New Zealand Economy, Working Paper, 03/03, New Zealand Treasury, Wellington, New Zealand.

Murray, W.E. (2006), *Geographies of Globalization*, Routledge, London.

Poot, J. (2004), 'Peripherality in the global economy' in Poot, J. (ed.), *On the Edge of the Global Economy*, Edward Elgar, Cheltenham, pp. 3–26.

Rabobank (2006) 'New Zealand dairy – is the honeymoon over?' *Rabobank Global Focus*, <www.rabobank.co.nz> (retrieved 5 December 2006).

Skilling, D. (2001), 'The Importance of Being Enormous: Towards an Understanding of the New Zealand Economy', Preliminary Draft – 30 July 2001, New Zealand Treasury, Wellington, New Zealand.

Skilling, D. and Boven, D. (2006a), *Developing Kiwi Global Champions: Growing Successful New Zealand Multinational Companies*, The New Zealand Institute, Auckland, New Zealand.

Skilling, D. and Boven, D. (2006b), *The Flight of the Kiwi: Going Global from the End of the World*, The New Zealand Institute, Auckland, New Zealand.

Stringleman, H. (2007), 'Fonterra angles for chance to build NZ's Nestlé or Nokia', *National Business Review*, 13 April.

The Economist (2000), 'Can the Kiwi economy fly?', 2 November 2000.

Van der Heyden, H. (2007), 'Fonterra's four-step plan to stay ahead of the world', *New Zealand Herald*, 9 April.

Waugh, J. (2004), 'WTO framework: Geneva talks revive dairy hopes', *New Zealand Dairy Exporter*, September, 72.

Yeung, H.W.-C. (2006), Situating regional development in the competitive dynamics of global production networks: an East Asian perspective, Plenary paper for the Regional Studies Association annual conference, London, 24 November 2006.

Chapter 16

The Commodity Chain at the Periphery: The Spar Trade of Northern New Zealand in the Early 19th Century

Michael Roche

Hopkins and I invented the term "commodity chain" to underlie a basic process of capitalism: that is it involved linked production processes that had always crossed multiple frontiers and that had always contained within them multiple modes of controlling labor.

(Wallerstein 2000, 221).

Introduction

Wallerstein's world systems theory offers a macro conceptualisation whereby it is possible to situate Aotearoa/New Zealand outside of the periphery of the world economy until first contacts from late 18th century but incorporated into it in the early 19th century and eventually by the 1890s, once the meat and dairy commodity chains develop, as part of the semi-periphery. In some assessments it had even assumed core status by the 1980s (e.g. Arrighi and Drangel in Terlouw 1992). Yet apart from once pondering whether it was 'possibly' part of the semi-periphery, Wallerstein (1976, 465) unsurprisingly says little about New Zealand and has written about geography only in terms of timespace (Wallerstein 1998). Even so Wallerstein's world systems theory had a resonance for geographers (e.g. Wallace 1990; Dodgshon 1993; 1998) and most notably provided the impetus for Peter Taylor's rethinking of political geography in the 1980s (Taylor and Flint 2000).

Commodity chains are an important part of Wallerstein's world system as the mechanism by which surplus value flows via a series of uneven exchanges from periphery to core. He has returned to commodity chains on several occasions (Hopkins and Wallerstein 1986; 1994; Wallerstein 2000a; 2000b; 2005). Other social scientists working on commodity chains typically identify Hopkins and Wallerstein (1986) as a foundational paper in the field and draw on its definition of a commodity chain (Hartwick 1998; Leslie and Reimer 1999; Bair and Ramsay 2003). Many then move to Gereffi and Korzeniewicz (1994) in extending their work (Daviron and Gibbon 2002; Bair and Ramsay 2003). Subsequently Bair (2005, 154) has argued that there is, however, a 'generally unacknowledged break' between the world systems and global commodity chain approaches. Global commodity chains themselves more recently have been subject to criticism by geographers interested in consumption rather than just production (Hughes and Reimer 2004).

This chapter revisits Wallerstein's commodity chain in the context of the shipbuilding an old value chain, but one that by the 1840s was to have its viability undermined by the transition from the age of sail to that of steam. It examines several nodes of this commodity chain as they related to New Zealand.

A Brief History of Wallerstein's Commodity Chains

The original Hopkins and Wallerstein (1986) paper was concerned not only with laying out the notion of a commodity chain but to support the view that a world economy was discernible by the late 18th century and that its origins could be traced back to the mid-17th century.[1] Hopkins and Wallerstein (2000, 223) noted, 'The concept of a "commodity chain" refers to a network of labor and production processes whose end result is a finished commodity'. They claimed to be making no assumptions about the geography of the chain, while simultaneously revealing trans-state divisions of labour.

In reconstructing commodity chains they stressed, 'we move backward rather than forward because we are interested in seeing the loci of the sources of value in a finished product and not the multiple uses to which raw material are put' (Hopkins and Wallerstein 2000, 224). Each major operation in the chain Wallerstein terms a 'node'. The simplest commodity chain identified inputs of tools, labour, and food to each node and the chain is formed by connecting nodes (Figure 16.1). This involved identifying the properties of each operation; Wallerstein suggested four elements:

- The usual nature of flows between node and operations that take place prior and subsequent to it.
- The dominant kinds of relations of production within the node.
- The dominant organisation of production including the technology and scale of unit of production.
- The geographical loci of the operation in question.

Wallerstein recognised that a commodity chain was dynamic and that by studying it over time he could look for major changes within a chain as well as the replacement of some segments by others. He suggested four signs of transformation within chains including, the geographical distribution of operations, the forms of labour encompassed by the chains, the technology and relations of production and the degree of dispersion/concentration of operations within each site of production.

Returning to commodity chains Wallerstein introduced a further element – 'boxes' (Hopkins and Wallerstein 1994, 18). Subsequently, he explained this was 'a

---

1 Wallerstein first published his ideas with Hopkins in 1986 in *Review* the journal of the Fernand Braudel Centre. It is reproduced more accessibly in his collection *The Essential Wallerstein* (2000), accordingly I have identified the 1986 paper but used the pagination from the collected volume.

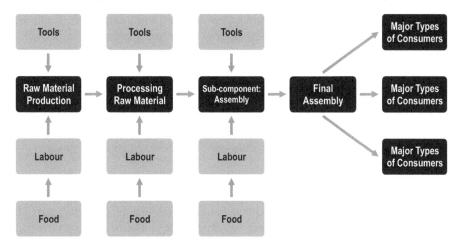

Figure 16.1   A simple commodity chain after Wallerstein

term chosen because it is without unnecessary connotations' (Wallerstein 2000a, 3). In elaboration he explained that:

> A box is thus a particular, quite specific production processes. The first thing to note about a box is that its boundaries are socially defined, and thus may be redefined. Boxes may be consolidated (where there were two). These redefinitions are effected through technological changes and/or social organisation changes.

(Hopkins and Wallerstein 1994, 18)

Focusing on a single box he posed six questions of any given commodity chain. First, to what extent was the box monopolised by a small number of units of production (i.e. is it core like with a high rate of profit)? Wallerstein suggested that for profitable boxes monopolies will attract competitors. Second, he asked, what is the geographic spread of the units of production? Thirdly, he posed a question about the number of different commodity chains in which the box was located suggesting that a number of outlets protected the producers' structure of production. Fourthly, he asked what kind of property-like arrangements are associated with units of production in the box, considering possibilities from petty producers, to larger entities private or parastalal, management by non owners, concession, and/or leases. Fifthly, he questioned whether there would be a sequence from various forms of wage employment to tenancy arrangements to non wage agreements and coerced labour. Sixthly, he inquired of the linkages joining the boxes in terms of purchase of inputs and sale of outputs. These questions are useful ones to bear in mind when considering accounts of the early spar trade in New Zealand.

Wallerstein's (2000a) engagement with commodity chains has involved some reiteration of his earlier ideas, but included further discussion of 'boxes' as well as what he termed 'single and multiple processes'. By the former he meant processes taking place at a particular locus usually involving a single firm. With multiple processes he complicated matters by not only allowing for multiple loci and firms but also suggested that 'multiple processes may take place in a single locus as well' (Wallerstein 2000a, 5).

While acknowledging a debt to Wallerstein and his commodity chain ideas geographers have tended to also see limitations. Thus Leslie and Reimer (1999, 404) regarded the global commodity chain literature as mainly deriving from a world system theory tradition so that, in the same way that world systems theory accounts are firmly located at the level of the modern world system, GCC [global commodity chain] narratives tend to centre upon the global dynamics of production/consumption/retailing linkages

In addition they note that 'the highly dualistic language of core and periphery is retained in a GCC approach' (Leslie and Reimer 1999, 404). They also highlight how under Wallerstein's approach of moving backwards from final production 'points of distribution and consumption are merely noted at the outset before commencing on the real task of unveiling production and extraction' (Leslie and Reimer 1999, 405). Many of those using global commodity chains acknowledge Wallerstein's early work but do not appear to have engaged with his later work (Wallerstein 1994; 2000a; 2000b; 2005). One exception is Daviron and Gibbon (2002, 139) who regard the use of the term 'box' as perpetuating a longstanding bias in work on vertical coordination which reduces it to the coordination of production via a set of technical stage and operations 'as opposed to the integration of different groups of agents'. More recently Bair (2005, 158) has considered how the world systems and global commodity chain literatures in their differing views on globalisation and objectives of analysis have further diverged with the latter moving away from 'the long-range historical and holistic analysis characteristic of the world-systems school towards networks and organisational approaches'.

The remainder of this chapter is concerned with the expansion of the shipbuilding commodity chain to Aotearoa/New Zealand in the early 19th century. This also provides a substantive instance, in which to rework Wallerstein's commodity chain ideas, especially where some of the conceptualisation has run ahead of empirical work.

The Shipbuilding Commodity Chain

Shipbuilding was identified for closer study by Hopkins and Wallerstein (2000, 226) on the grounds that 'ships constituted in this epoch the principal infrastructure for commodity exchanges as a well as an important locus of production (fish, whale oil, etc.)'. Subsequently Özveren (2000) identified several transformations in shipbuilding sites and supply zones from largely Mediterranean and NW Europe for the former and extending to Scandinavia for input supplies (exceptions being India and Cuba) in 1590 to Britain, Sweden and the USA by 1790. Özveren (2000, 62) also noted that by the late 18th century 'the shortage of masts had become sufficient to give rise to an ever-expanding seaborne commerce'.

Four 'sub-chains' comprised the shipbuilding commodity chain. These included the hemp sub-chain for sailcloth and rope, the ore sub-chains for ships fittings and armaments, the flax sub-chain for sail cloth, and lastly the timber sub-chain for spars and planking. Özveren's (2000) shipbuilding commodity chain is identical to that published by Hopkins and Wallerstein (1986). Yet Özveren (1994) had earlier produced a simplified version that subsumes masts within a timber sub-chain. Özveren's refers to timber and masts; the contemporary records and statistics for New Zealand use the term the spars. There is some ambiguity here in that the term is also used in a generic sense to include other sorts of ships timbers as well as spars for use in masts. I have followed the period usage of spar trade but tried to make it clear when other sorts of ships timbers were being secured from the forests of New Zealand.

Two of these sub-chains penetrated Aotearoa/New Zealand from the late 18th century; those for rope and timber, the former making use of New Zealand flax.[2] A shortage of ships timbers had posed major problems for Britain from the mid-16th century particularly after the loss of its North American colonies. From the 1790s small numbers of trees were felled for such purposes as an adjunct to sealing and flax gathering. The Royal Navy sent HMS Domendary in 1819 and HMS Buffalo, on two occasions in the 1830s to collect cargos of spars. Thereafter the Royal Navy secured supplies of spars from New Zealand by means of contracts. Albion (1926) notes that the contract system for spars created market like conditions and gave the Royal Navy a degree of quality control because they could specify delivery dates and reject as unsuitable on account of a single flawed spar entire consignments. A number of spar contracts were taken up including those awarded to Robert Brook who subcontracted to Sydney based trading firms such as Raine and Ramsey, those to local New Zealand traders such as Thomas McDonnell, Gordon Browne and the American William Webster. With colonisation in 1840 there were moves to establish forest reserves for masts and the short lived appointment of a conservator of kauri forests.

Masts from New Zealand, 1800–1840

The recent acquisition of the unpublished journals of Thomas Laslett who was on three early spar gathering missions to New Zealand adds significantly to an understanding the actual work in the forest of locating and preparing spars and their loading onto the ship (Colquhoun 2005). Laslett sailed on HMS Buffalo in 1834 and 1837–38 as second Purveyor of Timber and again in HMS Tortoise in 1839–41 as Purveyor of Timber. The British Admiralty required spars of quite precise dimensions from a minimum of 62ft [18.9m] and 20 inches [0.5m] diameter tapering to 11 inches [0.27m] up to 75ft [22.9m] and from 23 inches [0.6m] diameter to 13 inches [0.3m]. Kauri (*Agathis australis*) was the preferred species and they were intended for topmasts of 'line of battle ships'.

Laslett describes in some detail the search for trees of suitable dimensions, sometimes assisted by Maori guides, on other occasions exploring with members of the ship's party often hampered by swampy ground and thick bush. Once suitable trees were located in what Laslett termed a 'traverse' of the forest he marked them

---

2 *Phormium tenax* or New Zealand flax known to the Maori as Harakeke is related to the lily and is indigenous to Norfolk Island and New Zealand. It differs from the European flax of the genus Linum.

with the Kings's broad arrow.[3] The felling was done variously by groups of axe wielding Maori whose labour was provided by local chiefs in return for barter goods or carpenters and sailors from the ship. Frequently the instructions were misunderstood or disregarded as Maori cut the trees too short leading them to take up a cry of 'Taoyorta' – echoing the Timber Purveyors remark that it was 'too short' and thus useless as a mast.

What Laslett terms 'stages' were then built to provide a work platform from which the tree could be 'dressed'. The more elaborate ones were constructed out of timber sawn and on sloping ground could be 20–30 feet high [6.1–8.9m]. Preparation involved squaring of the butt of the log with cross cut saws and trimming off the branches. The bole was then squared by the carpenters using adzes. On numerous occasions a near completely shaped spar was found to be unsuitable because of defects such as the sapwood being too thick or a weakness in part of the trunk. While Laslett was unperturbed typically remarking only that much labour been expended on the spar, he noted that the Maori and sailors were apt to be discouraged. In addition to the spars other baulk and sawn timber was loaded to make up the cargo.[4] The contract spar gathers also had to accept a degree of wastage in the form of spars cut too short to be of use as part of the realities of dealing with a Maori work force.

Finished spars were either dragged by teams of Maori or sailors along prepared tracks of a mile (1600m) or more through the forest to the beach or occasionally to adjacent rivers and then floated to the coastal camp. Here they were stacked and then in small rafts of two of three towed out to the ship where they were hauled on board by the sailors using a pulley system. It took six to nine months to gather the hundred or so spars required. Progress was further delayed by wet weather and the need for Maori to leave to bring back additional food supplies for the forest parties. On other occasions, sailors, as opposed to the carpenters, were laid up with axe and adze cuts or otherwise injured in falls. These spar gathering ventures in New Zealand exemplify the global reach of the Royal Navy in attempting to secure its supply of masts. They reinforced the reputation of kauri and led onto a round of spar contracts that were filled largely by a number of New South Wales merchants who established stations in New Zealand.

---

3 Broad Arrow – three axe cuts in a tree truck; the stylised three feathers of the Prince of Wales symbol mark inscribed in the tree signifying that it was the property of the British monarch. Thomas MacDonnell did not hesitate to make use of the Broad Arrows mark in New Zealand in somewhat dubious circumstances.

4 Baulk is the term given to squared timbers of large dimension.

Table 16.1   Wallerstein's six questions of a box in a commodity chain with respect to the mast sub-chain in Aotearoa/New Zealand in the 1830s

Attributes	Characteristics
Number of units	Small number of spar producers limited by number of Admiralty spar contracts.
	Some entrepreneurs arrive e.g. McDonald.
Geographic spread of production	Limited by restricted bio-geographical range of Kauri.
	Search for and exploitation of available supplies.
	Spread of spar station throughout the Kauri forests districts.
Number of commodity chains	Admiralty spar gathering ventures also produced baulk timber, sawn timber and other sawn rough stowage.
	Merchants with spar contracts also produced sawn timber for export to NSW.
Property arrangements	Trees on Maori land bartered for directly between Europeans and chiefs.
	Merchants with spar contracts purchase trees from Maori and establish stations.
Employment types	Tribal labour made available by Chief giving way to barter moving for individual labour to wage arrangements for Maori labour.
	European labour always wage based (with some part payment in rum).
Linkages between boxes	Backwards integration by Sydney merchants owning the spar station.

Laslett's journals and other period accounts offer some substantive answers to Wallerstein's questions about 'boxes' applied to the mast chain (Table 16.1). Drawing on Laslett's journals it also possible to reconstruct the spar commodity chain from New Zealand in the 1830s (Figure 16.2). It differs from those laid out by Wallerstein and Hopkins, and Özveren reflecting the nature of the forests, the on-site processing, and the long voyage prior to final assembly. The shipbuilding chains set out by Wallerstein and his associates begins with 'planted forest' whereas the New Zealand kauri typically grew in patches in natural forest. Furthermore these forests were not free goods. They were in the possession of various Maori tribes and needed to be purchased. Initially this was by barter, for example in 1833 Captain Sadler of HMS Buffalo traded over 200 guns and ammunition plus quantities of blankets, fishing hooks, tobacco, iron pots and forks for his cargo of spars.[5]

_____

5 In detail this included; 20 seven barrelled guns; 10 single light ships fowling pieces; 200 muskets; 200 scabbards; 200 bayonets; 200 cartouch boxes; 200 frogs for bayonets; 500 flints for muskets; 100 flints for fowling pieces; 5000 cartridges ball; 500lb of shot lead for fowling pieces; 13lb of musket shot; 8 double barrelled fowling pieces; 8 bullet moulds; 40 bullet moulds for muskets; 5000lb of fine grain powder; 20 red and yellow jackets; 40 red and yellow trousers; 100 blankets; 4000 fishing hooks; 40 small blankets; 500 lb tobacco 50 cast iron pots; 150 three pronged forks (Capt. Sadler of the Buffalo to James Busby 11 December 1833, British Resident BR 1/1).

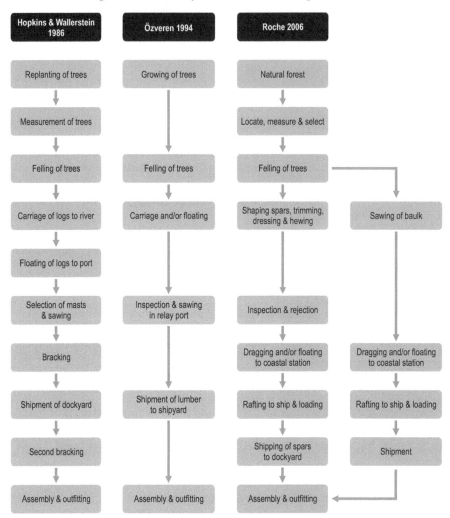

Figure 16.2   Mast commodity chains compared

Wallerstein (2000a, 9) suggests that 'As we move backwards in the chain, we are more likely to find one form or another of coerced labor. It is not however quite so simple a picture. For it is one that varies overtime'. Arguably Maori chiefs providing a workforce could be identified as coerced labour. Indeed Sadler negotiated the transaction with a small number of chiefs, but the barter goods were distributed to each of the men who laboured and Maori regularly left and returned with little explanation. One problem that emerged was the difficulty that the Navy officers had in explaining the precise dimensions of trees required. The sawyers in private spar gathering ventures understood the importance of the precise dimensions and freedom from flaws but many were also for a time paid partly in rum with the anticipated consequences as R.G. Jameson noted of William Webster's operation:

in the reckless and disorderly conduct of the Europeans in his employment, some of whom were the most persevering drunkards that it has ever been my fate to encounter ... To have withheld spirits from them would have determined them to seek other employment consequently he was absolutely compelled to have in his establishment a supply of rum, and to pay his people a greater portion of their wages in this (to them) irresistibly fascinating shape.

(Jameson 1842, 292).

Charles Terry commented on the rapidity with which the Maori recognised the value of both the trees and their labour,

For the service, formerly of very small remuneration to the Chief only, of a tribe was considered ample with a little tobacco among the whole of the labourers; but now the natives will not work, unless individually paid, and that at a high rate; and the European sawyers who formerly were glad to work at the rate of six shillings per 100 feet, now obtain the exorbitant rate of sixteen shillings and stipulate that the logs shall be placed on the pit for them.

(Terry 1842, 237).

The early spar trade warrants further attention; some of the ingredients are already available albeit in narrative form, for instance in Broeze's (1993) study of Robert Brooks an early to mid-19th century British ship owner and trader and in accounts of early shipbuilding in the south West Pacific (Hawkins 1971). The spar trade is also useful for rethinking Wallerstein's view that the commodity chain is constructed from the point of final consumption backwards to the raw materials. Those who secured the Admiralty spar contracts such as MacDonnell, Browne, and entrepreneurs such as William Webster were actively seeking to tie their part of the spar trade into the wider shipbuilding commodity chain. They were not passively waiting to be connected to or disconnected from the larger shipbuilding commodity chain centred on Europe.

Conclusion

Wallerstein's critics see world systems theory as introducing an unwanted core-periphery dichotomy into global commodity chain analysis though this is to overlook the conception of a semi-periphery. For geographers interested in consumption Wallerstein's commodity chains also harken back to an era of 'productionist' grand narratives. Remaining for the moment within such a frame of reference, I would observe that Wallerstein's 'vantage point' in discussing commodity chains is some where above the point of final assembly looking backwards to distant source areas. This vantage point was not shared by those producing spars in the forests of Aotearoa/ New Zealand and this has two implications:

1. For spar producers, their overview of the commodity chain was limited to perhaps only a few nodes. The ability to on sell the spars was all that really counted and they had some capacity to switch to sawn timber for sale in New South Wales and

elsewhere in Aotearoa/New Zealand as colonisation proceeded. Chain transformation was real enough, however as the development of steam ships effectively destroyed the spar trade by the mid-19th century.

2. A final assembly to production orientation means that is all too easy to see the spar gathers as passive hewers of wood and not as being quite active in trying to ensure their supplies of forest, procure labour, and secure further contracts.

Further work on the mast sub-chain of the shipbuilding commodity chain in Aotearoa/New Zealand in the 1830s is warranted not least because the voyage of the Buffalo was part of the Royal Navy's strategic search for suitable spars and not undertaken within a solely economic framework, so that the nodes that emerge from reading Laslett's journal may need to be modified in the light of the experience of the merchants' with spar contracts. Second there is the opportunity to look at how merchants attempted to ensure that they remained connected to the larger shipbuilding commodity chain.

Gereffi's has distinguished between buyer driven and producer driven commodity chains where Wallerstein's are of the former type; 'those industries in which large retailers, marketers, and branded manufacturers play the pivotal roles in setting up decentralised production networks in a variety of exporting countries, typically located in the Third World' (Gereffi 2001, 1620). Although Gereffi was not looking backwards to the 19th century it would be fruitful to rework buyer and producer led chains in this earlier temporal context.

Finally what does the world systems tradition of commodity chain research and the mast chain mean for agri-food studies? Hopkins and Wallerstein (1986) is still frequently cited by researchers including those working on the geographies of consumption. It has held less attention for geographers working on agri-food systems. Even researchers in the sociology of food and agriculture which has long standing interests in commodity chain analysis (e.g. Friedland 2001) have had only a limited dialogue with world systems commodity chain researchers.

The mast trade of the early 19th century may seem very far removed from the interests of 21st century agri-commodity chain research. Four points from this discussion on spars can be connected to a more general debate about agri-commodity chains:

1. Infinite regress in the construction of global commodity chains

Having constructed a simple commodity chain in conceptual terms Wallerstein acknowledged that it would be possible to depict chains for various inputs to that chain in a potential series of infinite regress, all of which arguably suggests that the term network or some such equivalent might be better than chain.

2. Consumption to original raw material production in commodity chain construction

Wallerstein has reconstructed buyer led commodity chains. However, to make this assumption for the convenience of constructing chains is to deny or at least diminish very active and creative efforts of traders at the periphery to create and connect segments of chains to larger commodity chains and to keep them connected. Dodgshon (1998) at least recognises the possibilities for innovation at the periphery.

3. Chains and sub-chains

The sailing ship commodity chain comprises at least five sub-chains with masts on occasions being given separate sub-chain status and at other times subsumed within the timber sub-chain. This raises questions about the definition of sub-chains and is also important when looking at production chains emanating from farms.

4. Bio products in commodity chains

Wood might be a relatively durable but it is still a natural product growing under a range of climatic and environmental conditions. This introduces an element of variability into the process that distinguishes it from more industrialised commodity chains. Not all kauri trees were suitable for spars, natural wastage could occur in harvesting and processing, specialists needed to inspect the finished products. Flaws that might not be immediately evident could cause the whole spar to be rejected. This situation impacted on the workforce reducing morale and subsequent productivity. These sorts of operating conditions have much in common with agri-food commodity chains.

For agri-food researchers, commodity chains as a concept whether constructed in Wallerstein's terms or more widely (Friedland 2001, 2004) need not be constrained to a world-systems framework. This makes it easier to reconstruct them other than solely from final assembly back to production. Work on historically significant commodity chains has hardly been exhaustive but is warranted to understand chain transformation; important for agri-food chains as well. The global commodity chain researchers might also consider if a less 'presentist' focus would assist them with their inquiries. Continuing to intersect commodity chains and globalisation research will also be fruitful as will further attention to Wallerstein's ideas about 'boxes' and 'single' and 'multiple' processes.

References

Albion, A. (1965), *Forests and Sea Power: The Timber Problem of the Royal Navy 1652–1862*, first published 1926, Hamden, Archon.

Bair, J. (2005), 'Global capitalism and commodity chains: looking back, going forward', *Competition and Change*, vol. 9, pp. 153–80.

Bair, J. and Ramsay, H. (2003), 'MNCs and global commodity chains: Implications for labor strategies', in Cooke, W.N. (ed.), *Multinational Companies and Global Human Resource Strategies*, Quorom Books, Westport.

British Resident BR 1/1, Archives New Zealand, Wellington, New Zealand.

Broeze, F. (1993), *Mr Brooks and the Australia Trade: Imperial Business in the Nineteenth Century*, Melbourne University Press, Melbourne.

Colquhoun, D. (2005), 'The New Zealand Journals of Thomas Laslett, 1833–1843', *The Turnbull Library Record*, vol. 38, pp. 85–91.

Cooke, W.N. (ed.) (2003), *Multinational Companies and Global Human Resource Strategies*, Quorom Books, Westport.

Daviron, B. and Gibbon, P. (2002), 'Global commodity chains and African export', *Agriculture Journal of Agrarian Change*, vol. 2, pp. 137–61.

Dodgshon, R. (1993), 'The early modern world-system: a critique of its inner dynamics', in Nitz, H-J. (ed.), *The Early Modern World-System in Geographical Perspective*, Verlag, Stuttgart.

Dodgshon, R. (1998), *Society in Time and Space: A Geographical Perspective on Change*, Cambridge University Press, Cambridge.

Friedland, W.H. (2001), 'Reprise on commodity systems methodology', *International Journal of the Sociology of Food and Agriculture*, vol. 9, pp. 82–103.

Friedland, W.H. (2004), 'Agrifood globalization and commodity systems', *International Journal of the Sociology of Food and Agriculture*, vol. 13, pp. 5–16.

Gereffi, G. (2001), 'Shifting strategies in global commodity chains, with special reference to the Internet', *American Behavioural Scientist*, vol. 44, pp. 1616–37.

Gereffi, G. and Korzeniewicz, M. (eds) (1994), *Commodity Chains and Global Capitalism*, Praeger, Westport.

Hartwick, E. (1998), 'Geographies of consumption: a commodity chain approach', *Environment and Planning D Society and Space*, vol. 16, pp. 423–37.

Hawkins, C.W. (1971), 'An introduction to European shipbuilding in the South West Pacific', *The Mariner's Mirror*, vol. 57, pp. 87–301.

Hopkins, T. and Wallerstein, I. (1986), 'Commodity chains in the world economy prior to 1800', *Review*, vol. 10, pp. 157–70.

Hopkins, T. and Wallerstein, I. (2000), 'Commodity chains in the world economy prior to 1800', in Wallerstein, I. (ed.), *The Essential Wallerstein*, New Press, New York, pp. 221–33.

Hopkins, T. and Wallerstein, I. (1994), 'Commodity chains: construct and research', in Gereffi, G. and Korzeniewicz, M. (eds), *Commodity Chains and Global Capitalism*, Praeger, Westport.

Hopkins, T. and Wallerstein, I. (1994), 'Conclusions about commodity chains', in Gereffi, G. and Korzeniewicz, M. (eds), *Commodity Chains and Global Capitalism*, Praeger, Westport.

Hughes, A. and Reimer, S. (2004), *Geographies of Commodity Chains*, Routledge, London.

Jameson, R.G. (1842), *New Zealand, South Australia and New South Wales*, Smith Elder, London.

Leslie, D. and Reimer, S. (1999), 'Spatialising commodity chains', *Progress in Human Geography*, vol. 23, pp. 401–20.

Nitz, H-J. (ed.) (1994), *The Early Modern World-System in Geographical Perspective*, Verlag, Stuttgart.

Özveren, E. (1994), 'The shipbuilding commodity chain 1590–1790' in Gereffi, G. and Korzeniewicz, M. (eds), *Commodity Chains and Global Capitalism*, Praeger, Westport.

Özveren, Y.E. (2000) 'Shipbuilding 1590–1790', *Review*, vol. 23, pp. 15–86.

Taylor, P. and Flint, C. (2000), *Political Geography, World-Economy Nation-State and Locality*, Prentice Hall, Harlow.

Terlouw, K. (1992), *The Regional Geography of the World-System*, Utrecht, Netherlands Geographical Studies, Rijksuniversiteit.

Terry, C. (1842), *New Zealand, Its Advantages and Prospects as a British Colony*, Boone, London.

Wallace, I. (1990), *The Global Economic System*, Unwin Hyman, London.

Wallerstein, I. (1976), 'Semi-peripheral countries and the contemporary world crisis', *Theory and Society*, vol. 3, pp. 461–83.

Wallerstein, I. (1991), 'The inventions of time space realities: towards an understanding of our historical systems', *Geography*, vol. 73, pp. 5–22.

Wallerstein, I. (1998), 'The time of space and the space of time: the future of social science', *Political Geography*, vol. 17, pp. 71–82.

Wallerstein, I. (2000), *The Essential Wallerstein*, New Press, New York.

Wallerstein, I. (2000a), 'Introduction', *Review*, vol. 23, pp. 1–13.

Wallerstein, I. (2000b), 'Next steps', *Review*, vol. 23, pp. 197–9.

Wallerstein, I. (2005), 'Protection Networks and the Commodity chains in the Capitalist World Economy', Paper presented at the Global Networks: Interdisciplinary Perspectives on Commodity Chains, 13–14 May 2005, Yale University, <http://fbc.binghamton.edu/iw-cc2005.htm>, retrieved 31 March 2006.

Chapter 17

Akaroa Cocksfoot: Examining the Supply Chain of a Defunct New Zealand Agricultural Export

Vaughan Wood

Introduction

For the last 130 years, commodities generated through pastoral farming have been the largest component in New Zealand's export trade. Indeed, for much of this time they have collectively accounted for 90 percent or more of it (Bloomfield 1984; Le Heron 1996). While many of these commodities, such as wool, mutton and dairy produce, have been highly visible in international markets, the trade in the raw material that has underpinned these commodity chains, that is grass seed, has attracted little attention. This may have been due to the highly localised nature of the industry in terms of seed production, together with the fact that grass seed, unlike most other pastoral commodities, is not a product for household or industrial consumption, but is instead absorbed back into the primary sector.

At first, the grass seed trade in New Zealand was essentially an import trade (see Stapledon 1928), but within a few decades of the start of organised European settlement in the 1840s, there were not only internal flows being generated, but also the direction of external flows was starting to reverse. The reason for this was that New Zealand was not the only country with an interest in transforming grasslands in the late nineteenth and early twentieth century. The pastoral frontier was also expanding in Australia, the United States, South America, and South Africa, and even where it was not, as in Great Britain, farmers were learning the value of imported seed stock for pasture renewal. With so much of New Zealand's intellectual and economic effort being concentrated on constructing introduced grasslands, it was perhaps inevitable that it would carve out for itself a leading place in world grass seed markets. For example, in the mid-1990s, New Zealand-grown seed accounted for more than half of the international trade in white clover seed (Clifford 1997).

The grass seed export New Zealand became renowned for in the late nineteenth century, however, was cocksfoot seed, or as Americans prefer to call it, orchardgrass seed. This was the first exotic commodity in which New Zealand gained global market dominance, and thus this industry was perhaps the nineteenth century equivalent of the kiwifruit industry. Its heyday, however, has long since past, and so this paper focuses on the years 1880–1940, when the industry was at its height. At this time, the major export destination was the United Kingdom, but as Figure 17.1

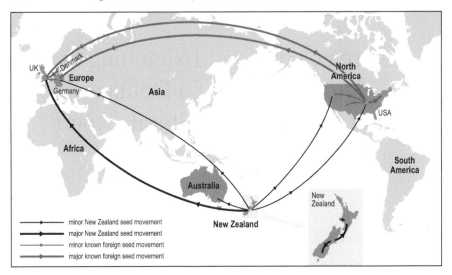

Figure 17.1   The global pattern of cocksfoot seed flows c. 1900–1910, showing main New Zealand seed and known foreign seed movements

Source: Based on contemporary New Zealand sales and export data (e.g. Canterbury N.Z. Seed Co. (1910) and New Zealand. Registrar-General (1911)), plus foreign comment on the cocksfoot trade (e.g. Hicks (1899) and Faber (1916)).

indicates, Australia, the United States, and Germany were also significant markets. In addition, there was also a considerable amount of demand from within New Zealand, particularly from the 'bush-burn' frontier in the North Island, and more latterly, from farmers trying to replace worn out indigenous grass pastures in the South Island.

To date, most commodity chain research has been concerned with studying globalisation processes, and thus its focus has been on 'standardised' products grown or made for the mass market, where competition is based on price. This has been contrasted with 'specialised' production, such as the produce of organic farms, where perception of quality is equally if not more important than price, and which tend to operate within a localised market (see Storper and Salais 1997; Murdoch *et al.* 2000). The New Zealand cocksfoot industry was an example of 'specialised' production, although it differs from many other 'specialised' industries in two respects. Firstly, unlike organic farming for instance, the barriers to entering and exiting the industry were very low, a characteristic which led the industry to take on a boom-and-bust type trajectory. Secondly, the focus of the market moved very quickly from being local to global. While it was local, seed merchants-cum-brokers who controlled both seed processing as well as sales were the driving force in the chain, but once it became global, foreign seed companies began conveying back to New Zealand the consumer expectations in terms of quality and price. The interdependence on each other, as the respective gatekeepers of supply and demand, meant that the export of cocksfoot seed from New Zealand became what has been termed a bipolar commodity chain (see Fold 2002; Ponte and Gibbon 2003).

Grass Seed as an Agro-Commodity

It is tempting to think that grass seed should be a commodity which is available to every pastoral farmer, and thus there is no need for a grass seed commodity chain. In practise, this is not the case, for in order to grow clean seed which can be sold to other farmers, it is imperative that the harvested area contain as few unwanted species as possible, which is best achieved from the growth of a species-specific monoculture.

A determined farmer could conceivably build up the seed needed for this monoculture by first collecting it from individual plants, but obviously economies of scale dictate that it will always be cheaper to purchase seed from the commercial supply. The grass-seed grower does nevertheless have one distinct advantage over growers of many other crops, namely that it is possible to switch the purpose of the crop at any stage from seed production to stock feed whenever prices or the condition of the crop suggests that the seed harvest will be an unremunerative one.

As there was natural variability in the output, it was difficult to guarantee the quantity and to some extent the quality of grass seed in any particular season. Quality was typically more important in the case of grass seed than in those commodities which served the household and industrial sectors, as the grass seed consumer was buying in new seed on an annual basis, whereas replacement of foodstuffs in the household and raw materials in industry was more or less continuous. Moreover, if seed used in permanent pasture was poor, its ill-effects would extend well beyond a year. Prior to the establishment of seed certification and testing regimes, which are essentially a twentieth century phenomenon (Copeland and Miller 1995; Great Britain Central Office of Information 1961), even competent buyers were not always able to tell at the point of a purchase how a consignment of seed would perform, or where the seed originated from, and thus what inherent characteristics it was likely to possess. To minimise this risk, buyers often sought out brands and seed merchants they knew to be reliable. In addition, they assessed quality on the basis of simple physical characteristics, such as seed weight, size and colour.

Grass seed was also a product which, like Christmas trees, could only be utilised by the consumer on a seasonal basis. If the seller missed the market in one year, they generally had to wait until the market become active again the following year before disposing of their stocks. This was a challenge New Zealand suppliers faced in the European and American markets, as these opened in the Northern Hemisphere autumn, following the local grass-seed harvest. Consequently, while Northern Hemisphere grass seed could move along the commodity chain without delay, there was an inherent six month time lag for seed coming from New Zealand (e.g. *Akaroa Mail* [hereafter *A.M.*] 15 July 1892). Where seed was being used for pasture renewal, as it was generally in the Europe, since the pastoral frontier there was already well established, there was also a high degree of elasticity in the market on a year to year basis, since this was an operation that farmers could easily defer. The instability in demand was a powerful disincentive to growers to rely on seed for their sole source of income, or make more than the minimum investment required for its production.

The Significance of Cocksfoot

On the basis of contemporary seed catalogues, it is clear that the New Zealand grass seed trade circa 1900 involved around 20 grass and 10 clover species (e.g. Yates 1904). Two species, however, stood out above the others, namely perennial ryegrass and cocksfoot, on account of their almost universal use in pasture seed mixtures. Together, for example, they accounted for more than £80,000 worth of trade for the Canterbury Seed Company in 1910, whereas combined sales for all other grasses and clovers amounted to less than £30,000 (Canterbury (N.Z.) Seed Co. 1910). Nationally, perennial ryegrass and cocksfoot seed production each accounted for 20,000–40,000 acres at this time (Evans 1956), whereas the combined area dedicated to grass and clover species of secondary importance in the seed industry, such as red and white clover, and Chewings' fescue, were much smaller; just before the First World War, for instance, the total for these other grasses and clovers was barely 20,000 acres (New Zealand Registrar-General 1912).

In the export trade, moreover, the pre-eminence of cocksfoot, and more particularly 'Akaroa cocksfoot', as Peninsula-grown cocksfoot was known, was unchallenged from the 1880s until the start of the First World War. New Zealand's export statistics do not discriminate between the various grasses, but we do know from where it was being shipped; as Figure 17.2 shows, the vast majority of it was leaving via Lyttelton, which is Canterbury's (and thus Banks Peninsula's) major

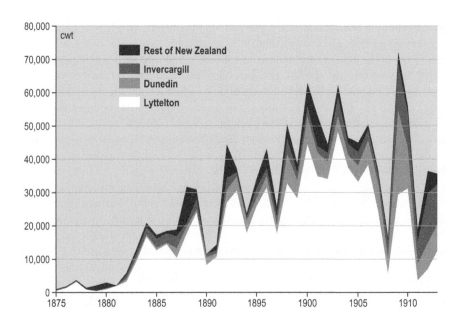

Figure 17.2   New Zealand grass seed exports, 1875–1913

Source: Based on data from annual volumes of the *New Zealand Gazette* and *Statistics of the Dominion of New Zealand.*

seaport. Written reports on the Canterbury seed industry in publications like the *New Zealand Country Journal* only refer to cocksfoot exporting, and during the 1880s published shipping returns for Lyttelton substituted the category 'grass seed' with 'cocksfoot' (e.g. *Lyttelton Time* 28 November 1888). On this basis, it can be inferred that most of the grass seed exported from New Zealand was cocksfoot seed. The inclusion of cocksfoot in the Lyttelton port returns with only six other commodities is also a testament to its importance in the Canterbury economy during these years.

The greater significance of cocksfoot, however, is reflected in the tremendous growth in exports which occurred during the 1880s. At that time, the suitability of perennial ryegrass in permanent pasture mixtures was being hotly debated in the United Kingdom, and some agricultural scientists were even suggesting that it should be excluded from them entirely. Furthermore, the grass most favoured by the ryegrass-critics to takes its place as the mainstay of permanent pastures was cocksfoot. The cocksfoot that grew on Banks Peninsula was particularly long-lived and leafy, which was exactly what these critics were looking for. As a consequence during the 1880s, New Zealand firms became the chief supplier in the much expanded British cocksfoot market (Stapledon 1931).

The Cocksfoot Production Process

One of the attributes that set Akaroa cocksfoot apart from many other agricultural crops was that it did not need regular cultivation as such. Its persistence in pasture meant that once the sward was established, it could be harvested for seed year after year without the need for re-sowing. As a consequence, growers needed to carry out only two operations per year, the first being the exclusion from September onwards of any cattle that might have been grazing on seed producing paddocks, and the second being the harvest itself, which normally took place over about six weeks in January and February (Stapledon 1927; Coulson 1979). Sheep grazing, it should be noted, was incompatible with seed production, as unlike cattle their bite is sufficiently close to the ground to damage the growing parts of the cocksfoot plant (White 1990).

When it came to harvesting, the growers of Banks Peninsula were assisted by a cast of thousands, made up primarily of itinerant labourers. Often they would be paid with a proportion of the crop. The reason the harvest was so labour-intensive was than much of it had to be carried out by hand, on account of the steep terrain across much of Banks Peninsula (the remnant of three volcanic cones, with the two principal harbours at Akaroa and Lyttelton being partially submerged calderas). After the crop had been cut with the sickle, the resulting stems were collected in bunches, and these were left to dry out. Threshing of the seed came next, and finally the seed was cleaned. Initially, the latter two operations were done by hand, using flails and riddles (large wooden sieves) respectively, but over time growers were able to take advantages of lightweight portable threshing and seed cleaning machines (Coulson 1979). During the 1880s and 1890s the harvest as a whole cost around 1.25–2d. per lb. of seed (*A.M.* 24 February 1885 and 6 May 1892; *Lyttelton Times* 14 September 1888). From a series of price ranges compiled by Moritzson, it can be discerned that the typical price paid in Christchurch for good quality riddled (that is, farmer-cleaned) seed was 2.5–3.5d per lb. (*A.M.* 28 January 1908).

With the harvest complete, the crop was either sold to a seed merchant, or consigned to a broker, who might auction it locally, or organise its export on the grower's account. Most of these seed merchants and brokerage firms were in Christchurch or Dunedin, so this required either sea or rail transport, although as these freight charges were relatively small, they would have had little effect on the price. If the seed was destined for export further mechanical cleaning was undertaken to give machine-dressed seed. The aforementioned price data compiled by Moritzson suggests that this usually earned a premium of about 1d. per lb. over riddled seed, but this was countered by the average 20 percent loss in quantity once the rubbish in the seed was removed (*A.M.* 28 January 1908). Finally, in shipping the seed to the UK, charges of about another 1d. per lb. were incurred (*A.M.* 24 February 1885). According to an Edinburgh seed merchant in 1891, most of this exported seed was sold by brokers, although British seed wholesalers also acquired seed by putting in orders with New Zealand seed merchants (*A.M.*, 24 February 1891). This did not necessarily mean sourcing seed through different New Zealand firms though, as there was an increasing tendency over time for them to adopt both trading and brokerage roles.

The Structure of the New Zealand Supply Chain

Relatively little is known about the growers of Banks Peninsula cocksfoot in a collective sense, as it was not until 1931, by which time the industry had undergone a marked decline, that they decided to form their own association (Coulson 1979, 74–9). This was able to enlist 173 growers by 1935, which compares with the figure of 313 growers given in 1908 by Adolph Moritzson, principal of one of the leading buying and brokerage firms (*Otago Witness* 11 March 1908; *A.M.* 10 December 1935). Some fourteen years earlier, Moritzson posted a copy of his firm's report of the cocksfoot crop to all 500 Banks Peninsula farmers, although it is doubtful that all of these were engaged in selling seed (*A.M.* 2 March 1894). While some of the delay in forming an association may have been due to farmers not identifying themselves primarily as cocksfoot growers, it probably also related to the difficulties met in traversing the Peninsula at earlier times; when, for instance, Moritzson conducted his first major business trip to the Peninsula in 1890, in which he talked to 258 growers, he thought it advisable, 'owing to the peculiar nature and difficulty of access of that country' to hire local guides, and even then it took him 'ten days' constant riding on horseback' to cover the whole district (*Otago Witness* 6 March 1890). Having said this, there were concerted attempts by growers in both 1900 and 1903 to form a joint-stock company, in order to reduce the likelihood of glutting the market with cheap seed. The 1900 bid failed, however, on the grounds that existing buyers could undermine it by offering a higher price, while in the latter attempt a draft prospectus was issued, but the company never registered, presumably due its attracting insufficient investment (*A.M.* 10 April 1900 and 23 June 1903). A schedule of voluntary production cuts put forward at the 1900 meeting indicates the scale of production amongst individual growers; large growers were deemed to be those with 300 acres or more in seed production, while small growers had less than 100 acres devoted to it.

In comparison, much more can be established about the mercantile side of the industry, in which firms could act as buyers, brokers, or seed cleaners. A further spin-off activity was seed cleaner manufacturing by agricultural engineering firms (see Coulson 1979). An examination of advertising in the Peninsula's newspaper, the *Akaroa Mail*, indicates that in the first years of the export industry, when the price of seed, as demonstrated by the values entered in the Lyttelton shipping returns, was high, there were a large number of buyers (at least ten in 1885). By the start of the 1890s, however, this number dropped to five or six regular buyers, with most of the firms initially involved having been forced out of the market. With one or two exceptions, these larger players, such as the Canterbury (N.Z.) Seed Co., Moritzson and Hopkin (which became A. Moritzson & Co), and Kaye and Carter, were still the main buyers in 1910. Another interesting feature of the market, and one which it had in common which the Australasian stock-and-station agency sector, was the tendency to vertical integration of services (Ville 2000); Moritzson and Hopkin, for example, was for several years only involved in brokerage, while its advertising in the *Akaroa Mail* suggests that the initial emphasis of the Canterbury (N.Z.) Seed Co. was in cleaning seed bought by other firms, rather than in buying seed itself. Such vertical integration was also common in the British agricultural seeds industry, as it allowed firms better control over the quality of the seed they were producing, and hence they could sell the idea of reliability to their customers; some, like Sutton & Sons and Arthur Yates, even went as far as growing the seed on their own farms. In-house growing was never attempted in New Zealand though, presumably because most growers regarded cocksfoot seed as a secondary crop. The introduction of complex machinery used to clean seed to the high grade needed for export (see *A.M.* 8 February 1895) probably contributed to the oligopolistic concentration of the industry as well. Only firms with access to a high throughput of seed could have afforded this equipment, which therefore served as a barrier to new players entering the market.

Where the cocksfoot seed industry differed from the stock-and-station agency sector was in its apparent lack of formalised connections to foreign companies. Only one regular buyer, New Zealand Loan and Mercantile Agency Co., had a British-based head office (see Ville 2000), and the evidence from the Canterbury (N.Z.) Seed Co. archives suggests that overseas supply contracts were not exclusive to particular companies either. In 1898, for example, orders were sent to the firms of David Allester and John Picard in London, and the previous year seed was also supplied to Ward & Co. in Bristol and Samuel McCausland & Co. in Belfast (Canterbury (N.Z.) Seed Co. 1897–9). Perhaps the best illustration of the *ad hoc* links between New Zealand and British firms though is the behaviour of Charles Lees, the Canterbury (N.Z.) Seed Co.'s manager in 1900. Lees had been gone to the UK on the company's behalf, but on his return proceeded to set himself up as an independent buyer (*A.M.* 5 January and 20 March 1900). It is possible that foreign seed merchants may have wanted to avoid carrying the overhead of buying and storing seed for export for several months before they could on-sell in their domestic markets; while New Zealand firms had to cope with overhead as well, in the meantime they were also deriving income by on-selling seed for local use. As the measures of seed quality during the early years of the industry were subjective, foreign seed merchants may also have been happier to leave seed selection to companies with local knowledge.

Information Flows and Quality Control

Two of the recurring themes in contemporary comment about Akaroa cocksfoot were flows of information about supply and demand, and quality control. This was because both had a significant influence on price. With respect to information flows, the original concern of both growers and buyers within New Zealand was to know the amount of cocksfoot seed that had been harvested from Banks Peninsula. After lobbying by the Akaroa Farmers' Club, production figures began to be recorded in the annual agricultural returns from 1885 onwards (*A.M.* 16 January 1885).

The most critical information needed by New Zealand growers, buyers and brokers, however, was the current price of cocksfoot in London (*A.M.* 24 February 1885). The New Zealand price could be influenced by news from London regarding the amount of unsold seed from the previous year and the size of the American crop, but equally, because New Zealand was the largest supplier, reports on the quantity and quality of the Banks Peninsula harvest could affect the London market. News of the bumper harvest in 1900, for example, immediately shaved 5s. per cwt. (about 10 percent) off the price that Akaroa cocksfoot could obtain (*A.M.* 10 April 1900). Large transactions in seed were thus highly sensitive, which led firms such as the Canterbury (N.Z.) Seed Co. to communicate with buyers via coded telegrams. Given that the structure of the industry meant that there was no route of direct communication between British seed merchants and the growers themselves, the latter had to rely on market reports emanating from firms like Moritzson and Hopkin (*A.M.* 22 March 1892). Growers were naturally suspicious of these reports though, especially when rival New Zealand brokers and buyers accused them of being bogus (e.g. *A.M.* 13 February 1891, 20 March 1894) Accordingly, growers campaigned for price information to be included in cables from the Agent-General in London, a move that the New Zealand government agreed to in 1898, although it was not put into effect until 1901 (*A.M.* 1 March 1898, and 26 April 1901).

During the period under study, the locus of quality control shifted three times between Europe and New Zealand. Initially, the only form of control on New Zealand seed appears to have been an indicative one, in the form of grading of seed by colour and weight (the optimum being bright seed weighing 18lb. per bushel). Consequently, at the start of the 1890s British seed merchants still recleaned most of the New Zealand seed they received (*A.M.* 24 February 1891). Nevertheless, having found that these firms shunned low quality crops, such as that from the 1884 harvest (*A.M.* 13 March 1885), New Zealand suppliers began taking more responsibility for the cleanliness of these export shipments. Indeed, it appears that by 1898 machine-dressing was effectively a pre-requisite, as in that year the cleaning plant operator C.L. Wheeler warned growers that shipping undressed seed 'would have the effect of neutralising all the good that has been done of late years to raise the standard value of Akaroa seed in London'. (*A.M.* 15 March 1898) New Zealand suppliers then employed brands to indicate to foreign buyers that this was high grade seed. (*A.M.* 20 March 1894; *Weekly Press* 6 September 1894)

All this changed in 1898, when Akaroa cocksfoot entered the modern world of seed testing. The responsibility for reporting on seed quality, and in particular, its germination rate, was given to Professor F.G. Stebler of the renowned Zurich testing station. In the years prior to 1912, this recorded average purity for New Zealand cocksfoot of 86 percent (which seems low until it is realised that immature cocksfoot seed was classed as an impurity), and a germination rate of 83 percent. (*A.M.* 31 March 1899; Cockayne 1912) From this time on, the New Zealand Department of Agriculture's seed-testing unit took charge of these analyses, which were carried out on numerous cocksfoot samples every year. Even so, these guarantees were only on selected samples, rather than on the crop as a whole. This difficulty was solved, however, when the Department initiated seed certification in 1932 (Wallace 1936). It was this development, in fact, that finally spurred the growers into forming their own aforementioned association.

The Displacement of Akaroa Cocksfoot from the Global Market

In considering the evolution of the international cocksfoot market in the late nineteenth and early twentieth century, it is important to recognise the distinctive demands on cocksfoot in different parts of the globe. At the risk of oversimplification, in the European one, which was the largest of the three main foreign markets, it was wanted for pasture renewal, whereas in the American and Australasian markets cocksfoot was used in pasture establishment. Where it was wanted for renewal, there was an option for buyers to defer purchase when the price was high, and thus the market was comparatively elastic; where the pastoral frontier was expanding in America and Australasia, on the other hand, farmers needed to sow cleared land immediately, and thus the demand was comparatively inelastic. In the main, the European market was fought for between American-grown cocksfoot and Akaroa cocksfoot, for while cocksfoot was also grown in France and Germany, the seed derived from both was not of great quality (*A.M.* 10 March 1891 and 15 July 1892). The American market was supplied largely from within (Hicks 1899), while in Australasia, the lower standard of seed demanded opened Banks Peninsula growers up to competition from those in New Zealand's North Island, and more latterly, from Australian growers as well (see *A.M.* 1–12 May 1896 passim.). It is also worth noting that the fallback position provided by the domestic American market served to limit the range of prices met with in Europe; whenever the Banks Peninsula crop was small, it was worthwhile for American firms to ship seed to Europe, whereas if the crop was large, they would keep their seed at home (*A.M.* 15 July 1892).

This equilibrium was upset, however, at the start of the 1910s, when Denmark began supplying significant quantities of seed to the European market (see Faber 1916). Denmark, of course, had no fallback domestic market, and thus its growers presumably had to sell in Europe for whatever price it could get. At first, this was not an alarming development for Banks Peninsula growers, as by this time they were relying more heavily on the Australasian market. The Canterbury (N.Z.) Seed Co., for instance, sold 64 percent of its cocksfoot seed to foreign buyers in 1898, but in 1910, only 31 percent of its seed went offshore. Nevertheless, it was a precarious situation to be in.

Then came the First World War which had a disastrous effect on the Banks Peninsula cocksfoot industry. The higher cost of labour during and after the war began to price Akaroa cocksfoot off the market, as whereas its harvest remained highly labour intensive, in the flat fields of the United States and Denmark, growers were able to utilise the full range of farm machinery (see Card 1923). At the same time, the pace of expansion of the pastoral frontier within New Zealand slowed markedly, and accordingly local demand began to drop off. Indeed, there was now much more emphasis on sowing temporary pastures on arable land. This was something that fast-growing Danish cocksfoot was better suited too, and thus by the mid-1920s New Zealand actually starting importing cocksfoot (Stapledon 1927; Canterbury Chamber of Commerce 1932). Even the New Zealand Department of Agriculture turned its back on Akaroa cocksfoot, testing Danish cocksfoot at an experimental farm near Ashburton, which was less than 100km from Banks Peninsula (Ward 1924). It should be said, however, that not all the factors behind the decline of Akaroa cocksfoot were related to changes in the seed industry. The high post-war prices for mutton and dairy produce encouraged both sheep farming and more intensive feeding of cattle on Banks Peninsula, both of which reduced its capacity for seed production (Canterbury Chamber of Commerce 1932). In addition, by this time, the drain on soil fertility from half a century of seed harvesting was becoming a significant problem (Coulson 1979). These difficulties notwithstanding, an attempt was made to revive the industry during the Depression years, an effort that was spearheaded by seed certification. This was too little, too late though, as the selection and certification of more persistent perennial ryegrass species in the meantime (Galbreath 1998), and the emergence of new commercial cocksfoot varieties in the United Kingdom, such as S. 37, partly bred from Akaroa cocksfoot (Welsh Plant Breeding Station 1933; Langer 1990), largely removed the problem which Akaroa cocksfoot had been used to solve. Consequently, production during the 1930s failed to reach half of its pre-war peak. The onset on the Second World War then finished off the decline that the First World War had started (see Coulson 1979).

Conclusion

From the beginning, the Akaroa cocksfoot industry had been based around delivering farmers seed of high quality rather than low price. Even so, it would have been difficult to ensure growers met rigorous quality standards, given they were typically growing cocksfoot to supplement their other farm income. This was in sharp contrast to the situation of the 'classical agro-commodity chains' examined by Gibbon (2001), where production of commodities such as cocoa and coffee in tropical colonies was controlled by state-run (or state-sanctioned) entities. Nevertheless, the Akaroa cocksfoot industry did share many of the features of these 'classical agro-commodity chains', such as the lack of dedicated supply contracts, and the centrality of quality conventions. As this paper has demonstrated, in the early years of the industry quality was indicated both by the market price and spatial branding (and, in convention theory terminology, 'market co-ordination' and 'domestic co-ordination') (Raikes *et al.* 2000). The latter was especially relevant, since the characteristic persistence of Akaroa cocksfoot was not available from other source areas of cocksfoot.

The importance of quality meant that the two places in the chain responsible for ensuring this, namely the New Zealand seed merchants who also offered seed cleaning, and the foreign seed merchants who guaranteed the quality to their customers, became the two poles of the bipolar commodity chain. Between the two poles there was a fine balance of power, with the foreign seed merchants being limited in the extent they could impose themselves as buyers on their New Zealand equivalents, since collectively the latter had a quasi-monopoly on what was perceived to be persistent cocksfoot. Both poles were also circumscribed in their actions though by the fact that in pasture improvement markets such as Europe, buyers could freely exit the market if price conditions were unfavourable to them, in the same way that growers could in New Zealand.

In the later years of the industry, however, there was a marked shift to judging the cocksfoot on its technical merits ('industrial co-ordination'). This meant that attributes could be readily tested for in a range of cocksfoot seed, and so relying on the name 'Akaroa cocksfoot' as a proxy for persistence was no longer necessary. This is in keeping with the trend seen in many commodity chains, where the institution of standards has allowed the outsourcing of production functions, since the standard acts as the guarantee of quality, rather than the oversight provided by in-house production (Ponte and Gibbon 2003). In Europe, the negative effects of such a shift on spatially-branded product have been countered by the introduction of *appellation d'origine contrôllée* classifications, such as champagne versus méthode champagnoise (Murdoch *et al.* 2000), but this approach would have been impossible for Akaroa cocksfoot as it was the foreign product in the main markets where it was facing competition. Ultimately, it is poignant to observe that the new British cultivar bred from Banks Peninsula cocksfoot seed retained its technocratic, and entirely aspatial name, S.37. Through this cultivar British farmers were able to retain the utility of Akaroa cocksfoot, long after the supply of the commodity itself had come to an end.

Acknowledgements

Vaughan Wood would like to thank Eric Pawson for his insightful comments and Marney Brosnan for her assistance preparing the illustrations. This research (part of the 'Empires of Grass' project, Principal Investigators Eric Pawson and Tom Brooking) was supported by funding from the Royal Society of New Zealand's Marsden Fund.

References

Bloomfield, G.T. (1984), *New Zealand, A Handbook of Historical Statistics*, G.K. Hall, Boston.

Canterbury (N.Z.) Seed Co. (1897–9, 1910), Sales Journal[s], Seed Co. archives, ARC 1991.89, Canterbury Museum, Canterbury, New Zealand.

Canterbury Chamber of Commerce (1932), Cocksfoot for Pasture and Seed, *Canterbury Chamber of Commerce Agricultural Bulletin 35*.

Card, D.G. (1923), *The Marketing of Kentucky Bluegrass and Orchard Grass Seeds*, Kentucky Agricultural Experiment Station, Bulletin 247.

Clifford, P.T.P. (1997), *Trifolium repens* L. (White Clover) in New Zealand, in Fairey D.T. and Hampton J.G. (eds), *Forage Seed Production, vol.1, Temperate Species*, CAB International, New York, pp. 385–93.

Cockayne, A.H. (1912), Seed Testing, *New Zealand Journal of Agriculture* vol. 5, no. 5, pp. 478–83.

Copeland, L.O. and McDonald, M.B. (1995), *Principles of Seed Science and Technology*, Chapman and Hall, New York.

Coulson, J.K. (1979), *Golden Harvest, Grass-Seeding Days on Banks Peninsula*, Dunmore Press, Palmerston North.

Evans, B.L. (1956), *Agricultural and Pastoral Statistics of New Zealand, 1861–1954*, Government Printer, Wellington.

Faber, H. (1916), Danish Grass Seed. Letter to the Ed. *The Times* [of London], 14 November 1916, 7.

Fold, N. (2002), 'Lead firms and competition in 'bi-polar' commodity chains, grinders and branders in the global cocoa-chocolate industry', *Journal of Agrarian Change*, vol. 2, no. 2, 228–47.

Galbreath, R. (1998), DSIR, *Making Science Work for New Zealand, Themes from the History of the Department of Scientific and Industrial Research, 1926–92*. Victoria University Press in association with the Historical Branch, Department of Internal Affairs, Wellington.

Gibbon, P. (2001), 'Agro-commodity chains: an introduction', *IDS Bulletin*, vol. 32, no. 3, pp. 60–8.

Great Britain Central Office for Information. Reference Division (1961), *Seed Improvement in Britain*, London, Her Majesty's Stationery Office.

Hicks, G.H. (1899), 'Grass seed and its impurities' in Hill, G.W. (ed.) *Yearbook of the United States Department of Agriculture 1898*, Government Printing Office, Washington D.C., pp. 473–94.

Langer, R. (1990), 'Pasture plants', in Langer R.H.M. (ed.) *Pastures, Their Ecology and Management* (2nd Edn.), Oxford University Press, Auckland, pp. 39–74.

Le Heron, R. (1996), 'Globalisation and the economy, evolving links and interactions', in Le Heron R. and Pawson E. (eds), *Changing Places, New Zealand in the Nineties*, Longman Paul, Auckland, pp. 27–43.

Murdoch, J., Marsden, M. and Banks, J. (2000), 'Quality, nature, and embeddedness, Some theoretical considerations in the context of the food sector', *Economic Geography*, vol. 76, no. 2, pp. 107–125.

New Zealand Registrar-General (1912), *Statistics of the Dominion of New Zealand for 1911*, Government Printer, Wellington.

New Zealand Registrar-General (1911), *Statistics of the Dominion of New Zealand for 1910*, Government Printer, Wellington.

Ponte, S. and Gibbon, P. (2003), 'Quality conventions and the governance of global value chains', paper prepared for the conference 'Conventions et Institutions, approfondissements théoriques et contributions au débat politique', 11–13 December 2003, Paris.

Raikes, P., Jensen, M.F., and Ponte, S. (2000), 'Global commodity chain analysis and the French *filiére* approach, comparison and critique', *Economy and Society*, vol. 29, no. 3, pp. 390–417.

Stapledon, R.G. (1927), 'Herbage seed production in New Zealand, III – cocksfoot', *Journal of the Ministry of Agriculture*, vol. 34, pp. 411–18.

Stapledon, R.G. (1928), *A Tour in Australia and New Zealand, Grassland and Other Studies*, Oxford University Press, London.

Stapledon, R.G. (1931), 'Seeds mixtures, controversies past, present, and future', in Hunter H. (ed.), *Bailliere's Encycopaedia of Scientific Agriculture*, Bailliere, Tindall & Cox, London, vol. 2, pp. 1104–18.

Storper, M. and Salais, R. (1997), *Worlds of Production, the Action Frameworks of the Economy*, Harvard University Press, Cambridge, Mass.

Ville, S. (2000), *The Rural Entrepreneurs, A History of the Stock and Station Industry in Australia and New Zealand*, Cambridge University Press, Cambridge.

Wallace, V.B. (1936), *Cocksfoot Seed Production, Ashburton County, 1935–36, A comparsion with Banks' Peninsula*, Canterbury Agricultural College, Christchurch.

Welsh Plant Breeding Station (1933), *An Account of the Organization and Work of the Station from its Foundation in April, 1919 to July, 1933*, Cambrian News (Aberystwyth) Ltd, Aberystwyth.

Yates, A. and Co. (1904), *Farm Seeds 1904*, Yates Records, MS 1706 Box 30, Auckland War Memorial Museum.

Ward, F.E. (1924), 'Ashburton Experimental Farm, notes on operations, season 1923–24, *New Zealand Journal of Agriculture*, vol. 29, no. 2, pp. 119–23.

White, J.G.H. (1990), 'Herbage Seed Production' in Langer R.H.M. (ed.), *Pastures, Their Ecology and Management* (2nd Edn.), Oxford University Press, Auckland, pp. 370–408.

Chapter 18

Biotic Exchange in an Imperial World: Developments in the Grass Seed Trade

Eric Pawson

Introduction

The establishment of new and improved grasslands was central to the trade prospects of the British Empire as well as a defining characteristic of its environmental transformation. 'In this colony we annually import an immense quantity of agricultural seeds' said the *Canterbury Times* in New Zealand on 20 November 1885. 'A new country begins of necessity by importing many commodities which require special skill in the production of them' (in Mackay 1887, 124–5). The increasing demand for wool, meat and dairy products in urban America, Britain and continental Europe drove the conversion of huge areas of forest, wetland and rangeland, in north and south America, and in Australasia, into 'empires of grass'.[1] By the 1920s it could be said that: 'Of every five-pound note that Great Britain expends on overseas products of any and every kind, twenty-five shillings goes in purchasing what is, in essence, worked-up grass' (in Stapledon 1928, v).

Grass seed was therefore one of the foundations of the expansion of nineteenth century international trade. Good seed, as the basis of pastoral agro-commodity chains, required – as the *Canterbury Times* said – 'special skill' in its production. It is the purpose of this chapter to discuss how this seed was developed in particular places and the extent to which it was exchanged over distance. It draws on the Latourian concept of 'centres of calculation', extending the usual application of this from public agencies such as botanic gardens to those commercial firms which sought to draw together plant materials and capture key characteristics in the form of improved seed. These characteristics, it is argued, were encapsulated in brands designed to facilitate the operation of 'networks of dissemination' through which seed was sold across the world. The argument is illustrated by a discussion of the operation of a small number of British and New Zealand seed companies between the 1850s and the 1920s. Towards the end of this period, increasing state regulation of seed production sought to generalise the calculative approaches to quality that such companies had pioneered during these years.

---

1 'Empires of grass' is the title of the project (03-UOO-036) supported by the Marsden Fund of the Royal Society of New Zealand from which this paper is drawn. The co-PIs are the author and Professor Tom Brooking, University of Otago.

Centres of Calculation

The concept of centres of calculation has proved a useful analytical tool in the current approach to understanding empire in terms of the dynamic webs of ideas, information, commodities and cultural norms that held together its different parts (Lester 2006). Latour's (1987) derivation of the concept used examples such as scientific societies, museums, and mapping agencies which gathered, coordinated and interpreted the information gleaned from voyages of exploration, cartography and specimen collecting. These sites acted as primary points where different materials were collected, or accumulated, often from a wide geographical reach, and made to work together by 'translation'. This is 'the process of relating different entities to achieve union and integration' (Barnes 2006, 151). Furthermore, it is 'a cycle of accumulation that allows a point to become a centre [of calculation] by acting at a distance on many other points' (Latour 1987, 222).

In this scheme, the key to evolving imperial knowledge was 'how to be familiar with things, people and events that are distant' (Latour 1987, 220). In order to collect information or items that both withstood retrieval across distance and were combinable through translation, it was necessary to ensure that collectors were trained and equipped. Hence the construction of systems of botanical classification, such as that of Linnaeus (Casid 2005). The map is another example, constructed from retrieval of measures of latitude, longitude and distance to render spatial knowledge transferable and recombinable. Thus 'the unifying feature of the work of the collector and the classifier is that they are both involved in generating a network of abstractions' (Miller 1996, 24). The collector is the explorer, the surveyor, the plantsman; the classifier is the cartographer, the botanist. In the Latourian scheme, they are distinct but not separate, united reflexively by the disciplines of expectation and training.

These ideas are readily applicable to the global networks of biotic exchange that characterised the European empires of the eighteenth and nineteenth centuries. Brockway's account of Kew is an early expression of the mobile and calculative characteristics of imperial relations. She described it as 'a control centre which regulated the flow of botanical information from the metropolis to the colonial satellites, and disseminated information emanating from them' (Brockway 1979, 7). The house, or 'scientific seat', of Sir Joseph Banks, President of the Royal Society from 1778 to 1820, has also been analysed as a centre of calculation (Miller 1996). Stewart (1999) has framed the public role of the coffee houses of early eighteenth century London in similar terms, explaining their importance in bringing circuits of natural philosophy and commerce together. By such means, then current ideas relevant to, for example, problems of navigation were disseminated.

Centres of calculation cannot therefore be considered apart from the networks in which they are embedded. They create knowledge-as-power; acting to draw places afar in together, both through accumulation and through dissemination. Mackay (1996) identified 126 collectors, located in parts of the world in which Britain had strategic imperial interests, who sent plant specimens to Banks. Besides ordering the collections along systematic lines, Banks also furthered the circuit of knowledge by donating these materials to institutions such as Kew and the British

Museum. Recent geographical work, drawing on Latour, has analysed the Royal Geographical Society as a centre of calculation (Driver 2001), and explored the role of the US Office of Strategic Services between 1941 and 1945 as a collector and producer of information for strategic military ends. Barnes notes that such centres 'also send reconstituted information out', thereby enabling them to effect action at a distance (2006, 156).

Networks of Dissemination

In these accounts, the actual mechanisms by which action is effected at a distance through dissemination are less clear than the ways in which cycles of accumulation and processes of calculation operate. Livingstone, in *Putting Science in its Place* (2003), adds valuable empirical weight to Latour's concepts, in his exploration of laboratories, museums, botanic gardens and hospitals. He has since extended this to examine the uneven ways in which new knowledge was read in particular places. Distant places were enrolled differentially in new knowledge networks depending on the usefulness of that knowledge. He shows how Darwin's work was received in specific ways in St Petersburg, Charleston and Wellington, reflecting the political contexts into which it was being read (Livingstone 2005). This interpretation extends that of Latour in respect of the fragility of knowledge, whilst supporting his point that the science is not 'universal', and does not 'spread "everywhere" without added cost'. In order to spread at all, however, it must look 'for the progressive extension of a network' (Latour 1987, 247, 249).

Most accounts focus on public or civic worlds of accumulation and calculation. But commercial firms could also act as centres of calculation, with their activities in extending networks of dissemination taking a particular form. The characteristics of products that have been assembled using specialist, calculative knowledge have frequently been distilled by means of branding. This has acted to transmit awareness of quality across geographical space. Brands emerged in the agro-food sector from the 1880s as valuable intangible assets for businesses extending their market reach through the use of new packing and transportation technologies (Wilkins 1992). A brand name and trademark enabled producers to identify and maintain contact with increasing numbers of consumers over greater geographical distances, by use of intermediaries and modern means of distribution. Brands also reduced transaction costs for consumers, permitting them to identify packaged and canned products of consistent quality (such Quaker Oats, Kelloggs, Campbells Soups and Heinz). Branding was becoming common amongst British brewers of bottled beer in the 1890s (Wilson 1994). Wilkins (1994, 26) observes that 'most of the food and beverage enterprises affected by the[se] new technologies quickly came to sell branded products'.

Branding was not only used by companies involved at the consumption end of agro-commodity chains. Seed companies also constructed brands, seeking to encapsulate their roles as centres of calculation in the development of plant materials, as well as to enable these materials to be transmitted over significant distances for use by primary producers. Their brands did, however, share characteristics with those companies which sold products such as cereal and soup to urban consumers. They employed widely appreciated cultural signifiers. Endorsements from the elite, particularly royalty, or from international exhibitions and shows, were used to convey associations of desirability as well as quality across great distances. Local endorsements were designed to ensure that the brand was 'read' in ways that would be understood in specific contexts. Company agents or branches fed back valuable information, furthering cycles of accumulation, about what worked and what would not in particular markets.

The Grass Seed Trade

The combination of perspectives on centres of calculation and networks of dissemination provides a means of framing an examination of the activities of seed companies working transnationally during the period before greater state involvement in seed production. It places emphasis on the calculative aspects of such exchange, opening a way out of the explanatory vacuum that characterises much writing on how networks of biotic exchange functioned. Influential accounts such as those of Crosby (1986) downplay the commercial, scientific or governmental frameworks within which plant exchange occurred. Often such exchange is described as serendipitous rather than the product of wider intentionalities. For example, William Pember Reeves' description of the arrival of weeds seeds in New Zealand: 'A colonial botanist has described how, in a certain seaport, he examined a heap of rubbish thrown on shore in the clearing out of ship's hold. On the rubbish-heap something like twenty specimens of strange weeds were growing' (Reeves 1902, 364). But far from being freeloaders, these weeds might have been the product of commercial design. One of the complaints of the Reading firm of Sutton & Sons in the 1860s was that the London cartel that had attempted to control the British seed trade up to that point dealt deliberately in adulteration. These was done by bulking out supplies with dead and weed seeds, and justified in terms of what customers had come to expect.

Sutton & Sons was one of the most prominent late Victorian seed houses, for which there is a significant (but far from complete) archive. Suttons developed a business model that emphasised quality control, using this to generate a considerable colonial trade. It dealt with customers in Australasia, for example, both by individual order and through local agents. It had some substantial competitors as an empire supplier, such as the London firm of Carters, but these have left little archival trace.[2] An exception is the Manchester firm of Arthur Yates, like Suttons originally established in the early nineteenth century. Yates set up an overseas branch in Auckland in the 1880s, and from there founded another in Sydney later in the same decade. By then, there were many New Zealand firms operating both locally and in the international arena, such as the Dunedin partnership of Moritzson and Hopkin. New Zealand had moved from being primarily a customer of British firms, to being

---

2 The analysis of Suttons activities is based mainly on records held by the Museum of English Rural Life, Reading University, which also has a small number of Carters catalogues (see also Corley 1994–97).

both customer and supplier of seed material. In the imperial seed trade, it came to occupy simultaneously a peripheral and a nodal position.[3]

The focus on firms rather than state agencies reflects the lack of formal state engagement with commercial biotic exchange in the nineteenth century, at least within British imperial networks. In New Zealand, Reeves commented on the rise in 'State energy' in respect of agriculture, observing that this was 'of but recent display'. Until the 1880s, the colony's 'huge pastoral estates were supposed to call for little supervision or support' (Reeves 1902, 359). This began to change with the drive in the 1880s and 1890s to add value to the products of grass, notably in the butter and cheese trades. Nonetheless, this did not extend at this time to serious state involvement in pasture improvement or quality control. In Britain and in New Zealand, there was for long no equivalent of the national seed testing stations first established in Germany and Denmark in the late 1860s, and operating in most European countries and in the USA by the end of the 1880s.

In 1901, the Departmental Committee appointed by the Board of Agriculture in London 'To Enquire Into The Conditions Under Which Agricultural Seeds Are At Present Sold' concluded that 'no widespread complaint of the quality of seeds sold throughout the country has been brought under their notice', but that 'one central seed-testing station' should be established 'under Government auspices' (1901, xiv). One of two dissenting opinions was from Leonard Sutton, one of the brothers in the partnership of Sutton & Sons. He saw no need for such a move, either for farmers or seedsmen. But by then Sutton & Sons had long been a centre of calculation in this respect.

Seed Companies as Centres of Calculation

Suttons was established in Reading, west of London, in the early nineteenth century, as a meal and flour sellers. It was Martin Hope Sutton, the founder's son, who turned the firm into a respected seed supplier in Britain and across the world. 'One of the earliest settlers in New Zealand', writing in the *Reading Mercury* (24 January 1885), claimed that 'Mr Sutton, Senior, shipped clover and rye grass seeds to me in the year 1841. Those seeds were above five months on the voyage, and were the first English grasses sown in the Auckland district of New Zealand. The Colony since then has made rapid strides, and the Reading seed firm have shipped seed for most of the rich pastures in New Zealand'. This declaration is at least evidence of the extent of Suttons' influence. The firm's *Farm Seed List* for 1865 states that they 'are constantly packing Farm and Garden Seeds for Australia, New Zealand, India, Africa and other Foreign Parts'; *Sutton's Autumnal Catalogue* for 1867 carries three endorsements from New Zealand, as well as two from India and four from the Cape of Good Hope. There is regular reference in the pages of New Zealand newspapers to confirm Suttons' significance. When an agricultural college was established at

---

3 The discussion about Yates draws on the archive in the Auckland War Memorial Musueum, and that of Moritzson and Hopkin from documentation in the Hocken Library, Dunedin.

Lincoln in the South Island in 1878, its first research report discussed grass plot experiments, 'the seed obtained from Messrs Suttons and Sons' (Ivey 1883, 7).

This reach reflects innovative dissemination practices adopted by the firm as early as 1840. Reading was on the Great Western Railway. Suttons used this advantage in introducing prepaid carriage of seeds to the home market, regardless of distance, and to extend their networks through publication of catalogues. The firm's good name is reflected in the volume of mail it received: about 1000–1500 letters daily in 1878. Martin Hope Sutton was an experimenter: he carried out thousands of trials on different soils to demonstrate 'the importance of laying land down to grass with the good varieties indigenous to each soil' (*The Times* 7 October 1901). He took a stand – noted in his many obituaries in 1901 – 'against the fraudulent systems then in vogue' (*British Journal of Commerce* November 1901), such as deficient production of seed, imperfect cleaning, the mixing of old seed with new, and adulteration of stocks with dead seed of other plants. He sought to circumvent the hegemony of the London Seed Association that was known for poor practices, by having seeds 'grown under his own superintendence, and from his own stocks, by men who would deliver the produce straight to [Reading] without the mediation of a wholesale house' (Cheales 1898, 15). In this way he resolved, in Latourian terms, problems of both collection and translation. He was also a strong supporter of the passage of The Adulteration of Seeds Act in 1869.

The Act did not eradicate poor practices, as a court case in 1877 demonstrated. In connection with this case, one of Suttons' rivals, the firm of James Carter and Co. wrote to *The Times* (7 December 1877) that 'The farmer need never fear getting adulterated seeds from respectable houses, as it is to the interest of the seedsman of reputation to exercise his technical knowledge for the benefit of his customer, otherwise his customer becomes dissatisfied, and when this is multiplied the seedsman loses his reputation'. Carters used an elaborate visual brand, with the byline 'Seeds For All Climates' above an engraved image of a sailing ship alongside a wharf on which are placed several boxes and barrels of Carters' grass and clover seeds. Carters also had their own seed farms, and carried the endorsement of both the Queen and the Prince of Wales. Likewise, Suttons had received the royal warrant, in 1858, and its catalogues bore the imprint 'Royal Berkshire Seed Establishment'. Its branding was even more striking.

The brand maintained common elements for use in different circumstances, including the visual grammar, but was adapted to the specifics of each market. The common elements were: the central image of Demeter, the goddess of corn, seeding the globe with 'Sutton's English Seeds for All Parts of the World'; the use of a registered trademark with royal endorsement; external measures of quality in the form of numbers of prize medals won; and the message 'Priced Lists Post Free to all parts of the World'. The version used in Britain indicated that the central image 'is a facsimile of the lid of Messrs Suttons special export boxes of seeds'; that for New Zealand, shown in Figure 18.1, carried written endorsements from the Hon John McLean of Oamaru and Mr W Faulkner of Wakefield, Nelson. There is no evidence that Suttons bought seed in New Zealand, and it seems to have sold by mail order until acquiring a colonial agent some time before the First World War.

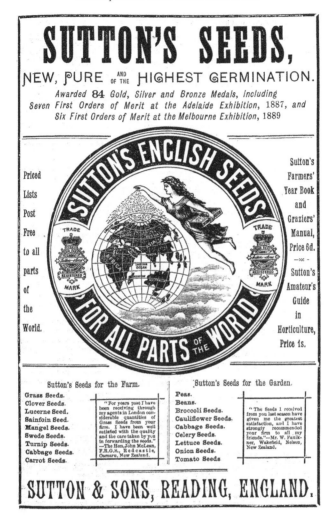

Figure 18.1 Sutton's brand

Source: Wise's Post Office Directory, 1890/91.

Arthur Yates & Co of Auckland claimed it was 'the only Colonial Seed Firm with headquarters in England'. The Auckland house was set up in 1883 by a grandson of George Yates, who had established a seed business in Manchester in 1826. After Arthur was joined by his brother in 1887, the former moved on to Sydney to start a branch there. The firm was known as Arthur Yates in both countries. It actively sought to protect its brand by warnings in its catalogues, as in the 1904 edition: 'In procuring our Seeds from Agents and Storekeepers, make sure our name is on the bags or tickets enclosed. We are repeatedly hearing of Seeds being sold as Yates' which were never supplied by us. There are cheaper, inferior samples to be had at a lower cost, but good, clean Seeds are always worth the money asked for them'. The firm's byline was 'Reliable Seeds'.

Like Suttons, it sourced its own, from its farms at Buckland in Auckland, at Exeter, New South Wales and from Manchester. It sought to make a point of distinction in the market with the statement 'Our Seeds are Re-cleaned [which] should be borne in mind when comparing our prices with others'. This is indicative of how the concerns of branded seedsmen had moved on, by 1900, from adulteration (against which a trademark might protect customers), to seed cleaning (a positive attribute of a brand). By 1924, its catalogues even carried photographs of an 'ordinary sample', of 'cleanings' and of a 're-cleaned' sample. As a centre of calculation, the firm asserted (in its 1904 catalogue) that 'We test the growth of every sample we sell, and do not send any seed out which does not germinate satisfactorily'. By then, this would have been standard practice amongst brand-name firms. Sixteen years earlier, the *Agricultural Gazette* had reported the laborious experimental process that Suttons used in 'determining the percentage of vitality in seed' (in *Lyttelton Times* 8 October 1888).

The growing importance of calculative practices is reflected in the activities and self-representation of a colonial seed house that purchased seed from New Zealand suppliers for the overseas market. Although firms such as Suttons and Carters could claim to have seeded the colonies, by the 1880s New Zealand was becoming a node for production of seed for use in Europe, North America and other colonies. This production often was not as tightly controlled as in the case of Suttons, Carters or Yates; usually colonial seed merchants purchased from individual farmers for whom seed was but one income stream. But some grasses grown in New Zealand acquired such high reputations, based on growth habit and palatability, that international purchasers were prepared to buy on this basis. A prime example was cocksfoot seed grown on Banks Peninsula, on the east coast of the South Island. In a printed circular to Peninsula farmers in 1891, the Dunedin agents Moritzson and Hopkin urged them 'to stencil your bags "Akaroa Seed"', this being the small settlement by which the district was most often signified. The nascent brand value rested on site of origin.

This approach was not calculative in the laboratory sense, but was in other ways. Moritzson and Hopkin promoted their role in the growing cocksfoot trade by collecting Peninsula seed output statistics and supplying prices in overseas markets. They quickly realised the depressive effect of contaminants in seed stocks, announcing in December 1894 that 'we are starting seed cleaning'. A month later: 'we shall strongly advise farmers to eradicate the Californian Thistle if it is growing on their land, as its presence in any seed makes the line almost unsaleable'. In 1900, they were 'strongly' advising farmers to send grass seed in for cleaning. 'Our plant is the best south of the line and is not surpassed anywhere'. By then the firm was giving annual reports on the seed harvest to growers with headings for 'exports', 'values', 'quality', 'colour', 'cleanliness' and 'germinating power'. In this way it was able to exert some authority with and over suppliers.

The examples of these firms are not, however, entirely typical of the means by which much seed was sourced. Grassland authorities were of the view that pasture establishment and management had for too long been taken for granted. The new director of Rothamsted research station in England observed that 'Notwithstanding its importance it cannot be said that grass receives the same attention as tillage crops ... it is far too generally considered that grass will grow without any assistance' (Hall 1903, 76). In New Zealand, A.H. Cockayne, the Biologist in the Department of Agriculture (established in 1892), estimated that weeds sown unintentionally in grass seed reduced the annual yield of pastureland by one million pounds each year. This he attributed to a still common wish for cheap seed: 'The desire for "bargains" in seed-buying is one that the farmer should sternly repress'. He urged them to use the department's free seed testing service or, in an attempt to improve self-regulation, its 'Weed-Seed Reference Card', supplied with a magnifying glass (Cockayne 1912, 4–5).

The clearest example of a state centre of calculation acting at a distance on a New Zealand site was that of the Welsh Plant Breeding Station. This was established at Aberystwyth in 1919 under Professor R.G. Stapledon, whose special interest was grassland improvement. Stapledon had been responsible for setting up what was to become the first British National Seeds Testing Station. In 1926–27, he visited Australia and New Zealand, and included Banks Peninsula on his route. He was aware of the good reputation of Akaroa cocksfoot, and had used it for seed development purposes (Stapledon 1927). Subsequently, he included it in new strains of cocksfoot that he developed at Aberystwyth in the 'S'-branded series of plants on which the station's name was made (Colyer 1982). Stapledon's careful experiments made available through networks of dissemination the best attributes of Akaroa seed for pasture production worldwide.

By now, his concern was not with the more readily calculable aspects of seed samples, such as cleanliness, germination percentage or price, but with strain, that is the suitability of plants for particular places and purposes. He decried the popularity after the First World War of cheap cocksfoot from Denmark, which was 'early, rather stemmy, broad tillered and lax' compared to the later developing, more leafy and palatable Akaroa variety. 'It is wholly to the disadvantage of the British farmer that Denmark [had by then] been allowed, to all intents and purposes, to completely replace New Zealand as a source of supply for the seed of such an important species as cocksfoot' (Stapledon 1927, 416, 418). Hence his development of the new S.37 and S. 143 strains from Akaroa material. The Welsh Plant Breeding Station was part funded by the Empire Marketing Board from 1928 to 1933, which helped to finance the scientific work done in New Zealand in the late 1920s by Stapledon's lieutenant, William Davies, who worked with Cockayne's grasslands acolyte, Bruce Levy. The work they did together was designed to be of long term benefit for pasture yields in both countries.

Conclusion

This chapter has sought to develop a relational framework to analyse aspects of global biotic exchange in the time from the mid-nineteenth century to the 1920s. This was a period when a number of companies emerged as significant players in the grass seed trade, developing an imperial reach based on calculative approaches to the creation and distribution of high quality product. By the early decades of the twentieth century, the encouragement of such approaches to pastoral commodity production was attracting greater state involvement. The analysis has applied insights from the literature on centres of calculation to commercial firms, identifying a range of assembly and networking practices that were used to derive and disseminate suitable materials for the establishment and improvement of far flung pastures.

The picture that emerges provides support for a dynamic understanding of imperial relations, based on flows of ideas, information and commodities. This is often crystallised in current writing on empire using the metaphor of the 'web' (Glaisyer 2004; Lester 2006). The links in the web were mobile and ever reformulating; fragile and in need of constant reinvestment; and, depending on the relation being described, cast particular places in different relations to each other. Thus a feature of the seed trade in the late nineteenth century was the manner in which peripheral places such as New Zealand's Banks Peninsula became for a while nodal suppliers of biotic material. In turn, the qualities of such material might later be incorporated in new cultivars for the improvement of the British pastures from which most colonial seed had originally been sourced. By the 1920s, the importance of the pastoral foundations of the imperial value chain in 'worked-up grass' was much more widely understood, due not least to the calculative improvements of grass seed by nineteenth century seed companies.

References

Barnes, T.J. (2006), 'Geographical intelligence: American geographers and research and analysis in the Office of Strategic Services 1941–1945', *Journal of Historical Geography*, vol. 32, pp. 149–68.

Board of Agriculture (1901), *Committee on Agricultural Seeds*, London, HMSO.

Brockway, L. (1979), *Science and Colonial Expansion, The Role of the British Royal Botanic Gardens*, Academic Press, New York.

Casid, J.H. (2005), *Sowing Empire, Landscape and Colonization*, University of Minnesota Press, Minneapolis.

Cheales, A.B. (1898), *M.H.S. A Few Noteworthy Incidents Connected with the Life of Martin Hope Sutton of Reading*, privately published, Reading.

Cockayne, A.H. (1912), 'Impure seed: the source of our weed problem', *Bulletin 21 NS*, Department of Agriculture, Commerce and Tourists, Wellington.

Colyer, R.J. (1982), *Man's Proper Study. A History of Agricultural Science Education in Aberystwyth 1878–1978*, Gomer Press, Llandysut.

Corley, T.A.B. (1994–97), 'A Berkshire entrepreneur makes good', *The Berkshire Archaeological Journal*, 75, 103–10.

Crosby, A.W. (1986), *Ecological Imperialism. The Biological Expansion of Europe*, Cambridge University Press, Cambridge.

Driver, F. (2001), *Geography Militant: Cultures of Exploration and Empire*, Blackwell, Oxford.

Glaisyer, N. (2004), 'Networking: trade and exchange in the eighteenth-century British Empire', *The Historical Journal*, vol. 47, pp. 451–76.

Hall, A.D. (1903), 'The manuring of grass lands', *Journal of the Royal Agricultural Society of England*, vol. 64, pp. 76–109.

Ivey, W. (1883), 'School of Agriculture, Lincoln, Canterbury', *New Zealand Country Journal*, vol. 7, pp. 46.

Latour, B. (1987), *Science in Action. How to Follow Scientists and Engineers Through Society*, Open University Press, Milton Keynes.

Lester, A. (2006), 'Imperial circuits and networks: geographies of the British Empire', *History Compass*, vol. 4, pp. 124–41.

Livingstone, D. (2003), *Putting Science in its Place. Geographies of Scientific Knowledge*, The University of Chicago Press, Chicago.

Livingstone, D. (2005), 'Science, text and space: thoughts on the geography of reading', *Transactions of the Institute of British Geographers*, NS 30, pp. 391–401.

Mackay, T. (1887), *A Manual of the Grasses and Forage-Plants Useful to New Zealand*, Government Printer, Wellington.

Mackay, D. (1996), 'Agents of empire: the Banksian collectors and evaluation of new lands', in Miller D.P. and Reill P.H. (eds), *Visions of Empire. Voyages, Botany, and Representations of Nature*, Cambridge University Press, Cambridge, pp. 38–57.

Miller, D.P. (1996), 'Joseph Banks, empire, and "centers of calculation" in late Hanoverian London', in Miller D.P. and Reill P.H. (eds), *Visions of Empire. Voyages, Botany, and Representations of Nature*, Cambridge University Press, pp. 21–37, Cambridge.

Reeves, W.P. (1902), *State Experiments in Australia and New Zealand*, George Allen & Unwin, London.

Stapledon, R.G. (1927), 'Herbage seed production in New Zealand: III – cocksfoot', *Journal of the Ministry of Agriculture*, vol. 34, pp. 411–18.

Stapledon, R.G. (1928), *A Tour in Australia and New Zealand. Grassland and Other Studies*, Oxford University Press, London.

Stewart, L. (1999), 'Other centres of calculation, or, where the Royal Society didn't count: commerce, coffee-houses and natural philosophy in early modern London', *British Journal of the History of Science*, vol. 32, pp. 133–53.

Wilkins, M. (1992), 'The neglected tangible asset: the influence of the trademark on the rise of the modern corporation', *Business History*, vol. 34, pp. 66–99.

Wilkins, M. (1994), 'When and why brand names in food and drink?', in Jones G. and Morgan N.J. (eds), *Adding Value. Brands and Marketing in Food and Drink*, Routledge, London, pp. 15–40.

Wilson, R.G. (1994) 'Selling beer in Victorian Britain', in Jones G. and Morgan N.J. (eds), *Adding Value, Brands and Marketing in Food and Drink*, Routledge, London, pp. 103–25.

Index